Studies in Economic Theory

Editors

Charalambos D. Aliprantis
Purdue University
Department of Economics
West Lafayette, IN 47907-2076
USA

Nicholas C. Yannelis
University of Illinois
Department of Economics
Champaign, IL 61820
USA

Titles in the Series

M. A. Khan and *N. C. Yannelis* (Eds.)
Equilibrium Theory
in Infinite Dimensional Spaces

C. D. Aliprantis, K. C. Border
and *W. A. J. Luxemburg* (Eds.)
Positive Operators, Riesz Spaces,
and Economics

D. G. Saari
Geometry of Voting

C. D. Aliprantis and *K. C. Border*
Infinite Dimensional Analysis

J.-P. Aubin
Dynamic Economic Theory

M. Kurz (Ed.)
Endogenous Economic Fluctuations

J.-F. Laslier
Tournament Solutions and Majority Voting

A. Alkan, C. D. Aliprantis and *N. C. Yannelis*
(Eds.)
Theory and Applications

J. C. Moore
Mathematical Methods
for Economic Theory 1

J. C. Moore
Mathematical Methods
for Economic Theory 2

M. Majumdar, T. Mitra and *K. Nishimura*
Optimization and Chaos

K. K. Sieberg
Criminal Dilemmas

M. Florenzano and *C. Le Van*
Finite Dimensional Convexity
and Optimization

K. Vind
Independence, Additivity, Uncertainty

T. Cason and *C. Noussair* (Eds.)
Advances in Experimental Markets

F. Aleskerov and *B. Monjardet*
Utility Maximization. Choice and Preference

N. Schofield
Mathematical Methods in Economics
and Social Choice

*C. D. Aliprantis, K. J. Arrow, P. Hammond,
F. Kubler, H.-M. Wu* and *N. C. Yannelis* (Eds.)
Assets, Beliefs, and Equilibria
in Economic Dynamics

D. Glycopantis and *N. C. Yannelis* (Eds.)
Differential Information Economies

*A. Citanna, J. Donaldson, H. M. Polemarchakis,
P. Siconolfi* and *S. E. Spear* (Eds.)
Essays in Dynamic
General Equilibrium Theory

M. Kaneko
Game Theory and Mutual Misunderstanding

Suren Basov

Multidimensional Screening

Dr. Suren Basov
University of Melbourne
Department of Economics
Melbourne, Victoria 3010
Australia
E-mail: s.basov@econ.unimelb.edu.au

Cataloging-in-Publication Data applied for

Library of Congress Control Number: 2004115347

Bibliographic information published by Die Deutsche Bibliothek
Die Deutsche Bibliothek lists this publication in the Deutsche Nationalbibliografie; detailed bibliographic data available in the internet at *http://dnb.ddb.de*

ISBN 3-540-23906-5 Springer Berlin Heidelberg New York

Springer is a part of Springer Science+Business Media
springeronline.com

Printed in Germany

Cover design: Erich Kirchner, Heidelberg
Production: Helmut Petri
Printing: betz-druck

SPIN 11356035 Printed on acid-free paper – 42/3130 – 5 4 3 2 1 0

I dedicate this book to my wife Svetlana and my daughter Margaret, with a provision that she will read it when she grows up.

Acknowledgement

This work was supported by the Melbourne University faculty research grant, FRG04. I am grateful to Peter Bardsley and Rabee Tourky for the comments on the first draft of the manuscript and students who took my Mathematical Economics class and spotted several typos in the text. All remaining mistakes are mine own.

Preface

In many industries the tariffs are not strictly proportional to the quantity purchased, i. e, they are nonlinear. Examples of nonlinear tariffs include railroad and electricity schedules and rental rates for durable goods and space. The major justification for the nonlinear pricing is the existence of private information on the side of consumers. In the early papers on the subject, private information was captured either by assuming a finite number of types (e. g. Adams and Yellen, 1976) or by a unidimensional continuum of types (Mussa and Rosen, 1978). Economics of the unidimensional problems is by now well understood.

The unidimensional models, however, do not cover all the situations of practical interest. Indeed, often the nonlinear tariffs specify the payment as a function of a variety of characteristics. For example, railroad tariffs specify charges based on weight, volume, and distance of each shipment. Different customers may value each of these characteristics differently, hence the customer's type will not in general be captured by a unidimensional characteristic and a problem of multidimensional screening arises. In such models the consumer's private information (her type) is captured by an $m-$dimensional vector, while the good produced by the monopolist has n quality dimensions.

Recent years witnessed a considerable progress in the area of multidimensional screening. The most important papers are Armstrong (1996), Rochet and Chone (1998), and Basov (2001, 2002, 2004). multidimensional screening models can have fruitful application in the theory of industrial organization, labor markets, and the optimal taxation. For an example of the latter, see Shapiro (2001). However, despite the wide prevalence of nonlinear tariffs and practical importance of understanding the economics that underlies them, this literature remains unknown to the most economists, beyond a circle of a few specialists in the area. In my view, the main obstacle to understanding of this literature is its use of mathematical tools, which are not a part of a standard economist's toolbox. This book attempts to remedy the situation.

I assume that the reader is familiar with basic mathematical tools used by in modern economic theory, such as calculus, linear algebra, elements of the theory of ordinary differential equations, the basics of the measure theory (so I use without a definition such terms as a Borel set, Lebesgue measure, or a Lebesgue measurable function), and the optimal control. I also assume knowledge of microeconomics at the level of Mas-Colell, Whinston, and Green (1995). However, I introduce all more advanced techniques.

The first part of this book contains a review of vector calculus, theory of partial differential equation of the first and second order, and the theory of generalized convexity. These techniques are extensively used in the multidimensional screening models. It also contains a chapter of miscellaneous techniques, which are some times useful in screening models, but are not used as extensively. When the results can be found in easily accessible literature I usually formulate them without a proof, referring the reader to another source. I, however, provide more thorough discussion of the results that are harder to access in the literature.

Part two is devoted to the economics of screening models. It starts with a detailed discussion of economics and mathematics of unidimensional screening problems and three approaches to their solution: direct, dual, and Hamiltonian. It uses the Hamiltonian approach to unify all known results, which were previously obtained using different arguments.

After a thorough discussion of the unidimensional case I move on to the multidimensional screening model. First, I discuss the main economic constraint, which makes the multidimensional case qualitatively different from the unidimensional one: path independence of information rents. I show that it significantly restricts usefulness of the direct approach in the multidimensional case. Dual approach can be taken somewhat further, but it also breaks down, when the dimensionality of type and the number of the screening instruments differ. Hamiltonian approach, however, can be applied fruitfully even in this case. For the most part of the book I, following almost all the literature in the area, assume that the consumer's utility is quasilinear in money. However, in Chapter 8 I go beyond the quasilinear case and provide some results, which apply to a more general specification. I also argue that is some applied models that arise the most natural specification is not quasilinear.

The final chapter concludes and outlines some directions for the future research. One of them, already mentioned, is going beyond the quasilinear case. Another is to look at screening models with finitely many goods and an infinitely dimensional type. I argue that this is probably the most relevant case from the economic point of view and give some thoughts how the techniques developed in the book can be adapted to handle it.

Each chapter, except for the last one, is followed by exercises. That makes the book useful for instructors, who wish to use it to teach an advanced graduate course in multidimensional mechanism design. Exercises are also important for the general reader, since they allow to develop mathematical and economic intuition and provide deeper understanding of the theoretical facts. Though this book should be sufficient to teach the reader all main mathematical techniques in the area of the multidimensional screening, I provided bibliographic notes in the end of each chapter, so an interested reader can achieve a deeper understanding of the material covered in the chapter and investigate the connections with a broader mechanism design perspective.

I hope that collecting necessary technical tools and economic insights of the multidimensional screening literature in one volume will facilitate the use of the results by applied economists and help to shed some light on many areas of economic theory.

Melbourne
October, 2004

SUREN BASOV

Contents

Part I

Mathematical Preliminaries

This part reviews the mathematical techniques that will be used in the second part of the book but do not usually form a part of a toolbox of an economic theorist. They include vector calculus with the elements of field theory, theory of partial differential equations of the first and second order, elements of the theory of Lie groups, the theory of generalized convexity, and the calculus of variations. The theory is supplemented by abundant examples and exercises. In this book I also use a convention that zero is a natural number.

1

Vector Calculus

This chapter contains a review of vector calculus and introduces the basic operations, such as the divergence, the gradient, and the Laplacian. Since all this material is well-known, I will give the proofs only if they are simple. For more complicated proofs I will refer the reader to the easily accessible standard text by Smirnov (1964). I will, however, supply the basic intuition behind the main results and explain their meaning.

In this book I will use the following notational conventions: (x, y) denote coordinates on a plane, R^2, (x, y, z) denote coordinates in R^3, $(x_1, ..., x_n)$ denote coordinates in R^n for $n > 3$ or a generic n. The bold letters always denote vectors in R^n. For any open Ω, I denote by $C^k(\Omega)$ the set of real-valued k times continuously differentiable functions. I also refer to functions of class $C^1(\Omega)$ as smooth functions.

1.1 The Main Operations of Vector Calculus: *div*, *grad*, and Δ

In multidimensional screening models the type of a consumer is usually parametrized by a *vector*. Some variables of economic interest, such as allocations of goods, are also vectors, while others, such like consumer surplus, are scalars. An important tool in dealing with such objects is vector calculus. Let us start with the following definition.

Definition 1 *Let $\Omega \subset R^n$ be an open, connected set. A function $\mathbf{a} : \Omega \to R^n$ is called a vector field on Ω. If each component $a_i \in C^k(\Omega)$ the vector field is called k times continuously differentiable.*

For example, if Ω is the set of possible types of consumers, an infinitely differentiable allocation of n goods will be a vector field on Ω. For any $x \in \Omega$, $\mathbf{a}(\mathbf{x})$ is a vector with components $(a_1(\mathbf{x}), ..., a_n(\mathbf{x}))$. An important operation on smooth vector fields is the *divergence*, defined by

$$div(a) = \sum_{i=1}^{n} \frac{\partial a_i}{\partial x_i}. \tag{1.1}$$

We will see below that the divergence can be associated with the density of sources of the vector field.

Definition 2 *Let $\Omega \subset R^n$ be an open, connected set. A function $f : \Omega \to R$ is called a scalar field on Ω. If $f \in C^k(\Omega)$ the scalar field is called k times continuously differentiable.*

A leading example of a scalar field in a screening problem is the consumer's surplus as a function of type. An important operation on smooth scalar fields is gradient, which is defined by:

$$gradf \equiv \nabla f = (\frac{\partial f}{\partial x_1}, ..., \frac{\partial f}{\partial x_n}). \tag{1.2}$$

Note that while operation of divergence maps vector fields into scalar fields, operation of gradient maps scalar fields into vector fields.

If $f \in C^2(\Omega)$ one can combine the operations of divergence and gradient to obtain a new operation, the Laplacian, defined by

$$\Delta f = div(\nabla f) = \sum_{i=1}^{n} \frac{\partial^2 f}{\partial x_i^2} \tag{1.3}$$

Note that Laplacian takes a scalar field into a scalar field. Laplacian of a vector field is a vector field, whose components are equal to the Laplacian of the coordinates of the original field.

In economic applications the variables of interest might fail to be sufficiently smooth on the entire set Ω. In Section 5.1 I will define generalized derivatives, which will allow us to differentiate in an appropriate sense any measurable function. All operations of the vector calculus will be generalized accordingly.

1.2 Conservative Vector Fields

Consider a vector field $\mathbf{a}$ defined over an open set $\Omega \subset R^n$. Let us establish the conditions that guarantee that there exists a scalar field φ such that

$$\mathbf{a} = \nabla\varphi. \tag{1.4}$$

If such a φ exists we will call it a *potential* of vector field $\mathbf{a}$ and vector field $\mathbf{a}$ is called a *conservative vector field.* The name *conservative* comes from physics. There if a particle moves in a conservative force field its mechanic energy is conserved.

It is obvious that if φ is a potential of vector field $\mathbf{a}$, so is $\widetilde{\varphi} = \varphi + C$ for an arbitrary constant C. Moreover, it can be shown that any potential of $\mathbf{a}$ has this form. To understand the issues involved in the existence of a potential let us start with some examples. Let $\Omega \subset R^2$ and consider a vector field $\mathbf{a}_1 = (x, y)$. Then $\mathbf{a}_1$ satisfies (1.4) for

$$\varphi = \frac{1}{2}(x^2 + y^2). \tag{1.5}$$

Now consider a vector field $\mathbf{a}_2 = (y, 2x)$. I claim that this vector field is not potential. To see why, assume it has a potential $\Psi(x, y)$. Then according to (1.4) we can write

$$\begin{cases} \partial\Psi/\partial x = y \\ \partial\Psi/\partial y = 2x \end{cases}. \tag{1.6}$$

Differentiating the first of the equations of system (1.6) with respect to y and the second with respect to x, one obtains that

$$\partial^2\Psi/\partial y\partial x = 1 = \partial^2\Psi/\partial x\partial y = 2. \tag{1.7}$$

But the last equality contradicts the Young's theorem. The observations made so far can be summarized in the following theorem:

Theorem 3 *Let* $\mathbf{a}$ *be a vector field defined over convex open set* $\Omega \subset R^n$. *Then* $\mathbf{a}$ *is conservative if and only if*

$$\frac{\partial a_i}{\partial x_j} = \frac{\partial a_j}{\partial x_i} \tag{1.8}$$

for all i *and* j.

Proof. To prove the necessity, assume that there exists a scalar field φ such that (1.4) holds. Then

$$\frac{\partial a_i}{\partial x_j} = \frac{\partial^2\varphi}{\partial x_j\partial x_i}, \tag{1.9}$$

while

$$\frac{\partial a_j}{\partial x_i} = \frac{\partial^2 \varphi}{\partial x_i \partial x_j}, \tag{1.10}$$

and (1.8) follows from the Young's theorem.

I will prove sufficiency only for $n = 2$. The main idea of the proof remains the same for arbitrary n (see, for example, Smirnov, 1964). In the case $n = 2$ assume that (1.8) holds and define function φ by

$$\varphi(x, y) = \int_{x_0}^{x} a_1(t, y)dt + g(y), \tag{1.11}$$

where $(x_0, y) \in \Omega$ for some y and $g(\cdot)$ is some continuously differentiable function. (Note that since Ω is assumed to be convex, $a_1(t, y)$ is defined on the segment connecting (x_0, y) with (x, y)). It is clear from (1.11) that

$$\frac{\partial \varphi(x, y)}{\partial x} = a_1(x, y). \tag{1.12}$$

I will show that one can select $g(\cdot)$ in such a way that

$$\frac{\partial \varphi(x, y)}{\partial y} = a_2(x, y). \tag{1.13}$$

For this purpose, differentiate equation (1.11) with respect to y to obtain

$$\frac{\partial \varphi(x, y)}{\partial y} = \int_{x_0}^{x} \frac{\partial a_1(t, y)}{\partial y} dt + g'(y) \tag{1.14}$$

and equate it to $a_2(x, y)$. Taking into account (1.8) we can evaluate the integral

$$\int_{x_0}^{x} \frac{\partial a_1(t, y)}{\partial y} dt = \int_{x_0}^{x} \frac{\partial a_2(t, y)}{\partial x} dt = a_2(x, y) - a_2(x_0, y). \tag{1.15}$$

As a result we get the following differential equation for $g(\cdot)$

$$g'(y) = a_2(x_0, y). \tag{1.16}$$

Solving it and substituting the result into (1.11) one obtains:

$$\varphi(x, y) = \int_{x_0}^{x} a_1(t, y)dt + \int_{y_0}^{y} a_2(x_0, z)dz + C, \tag{1.17}$$

where y_0 is such $(x_0, y_0) \in \Omega$ and C is arbitrary constant. Therefore, if (1.8) is satisfied, vector field **a** admits a potential. Moreover, the potential is uniquely defined up to an additive constant. ■

1.3 Curvilinear Integrals and the Potential

The importance of conservative vector fields in multidimensional screening theory comes from the geometric meaning of the existence of a potential. It turns out that the integral of such vector field between any two points does not depend on the path taken. We will see in the second part of the book that this property of conservative vector fields lies behind the possibility to define an unambiguous information rent in multidimensional models.

To formalize these ideas let us the following definitions.

Definition 4 *A smooth curve Γ is a continuously differentiable mapping $\gamma : [0,1] \to R^n$.*

Definition 5 *Let $\mathbf{a}$ be a vector field defined over convex open set $\Omega \subset R^n$. A curvilinear integral of the second type of the vector field along the curve γ is*

$$\int_{\Gamma} (\mathbf{a}, \mathbf{dx}) \equiv \int_0^1 \sum_{i=1}^{n} a_i(\gamma(t))\gamma_i'(t)dt. \tag{1.18}$$

In general the curvilinear integral depends on both: endpoints and the path which connects them.

(The curvilinear integral of the vector field of the first type is defined to be a vector

$$\int_0^1 \mathbf{a}(\gamma(t))\sqrt{\sum_{i=1}^{n} \gamma_i'^2(t)dt}. \tag{1.19}$$

I will not use the integrals of the first type in this book. Therefore, wherever I refer to a curvilinear integral I always mean the integral of the second type).

In general a curvilinear integral between two points depends on a path, connecting them. To see this, consider the following example.

Example 6 *Let $\mathbf{a} = (y, -x)$ and curves Γ^1 and Γ^2 are given by $\gamma_1^1(t) = t$, $\gamma_2^1(t) = t$ and $\gamma_1^2(t) = t$, $\gamma_2^2(t) = t^2$ respectively. Note that both curves connect the same points $(0,0)$ and $(1,1)$. Let us evaluate the curvilinear integrals of $\mathbf{a}$ along both of these curves.*

$$\int_{\Gamma^1} (\mathbf{a}, \mathbf{dx}) = \int_0^1 (t - t)dt = 0 \tag{1.20}$$

$$\int_{\Gamma^2} (\mathbf{a}, \mathbf{dx}) = \int_0^1 (t^2 - 2t^2)dt = -\frac{1}{3}. \tag{1.21}$$

Therefore, the values of curvilinear integrals along different paths connecting the same points differ.

It is straightforward to check that the vector field $\mathbf{a}$ from the previous example is not conservative. It turns out that if a vector field is conservative the curvilinear integral between two points does *not* depend on the path of integration.

Theorem 7 *Let* $\mathbf{a}$ *be a conservative vector field with potential* φ *and let* Γ *be any smooth curve connecting points* A *and* B $(\gamma(0) = A,\ \gamma(1) = B)$*. Then*

$$\int_\Gamma (\mathbf{a}, \mathbf{dx}) \equiv \varphi(B) - \varphi(A). \tag{1.22}$$

In particular, the value of the curvilinear integral depends only on the start and the end points, but not on the path taken. Vice versa, if for any two smooth paths connecting any two points in Ω *the value of the integral is the same it admits a potential.*

I will not give here a proof of this Theorem. An interested reader is referred to Smirnov (1964). Instead, let us consider the following example.

Example 8 *Let* $\mathbf{a} = (y, x)$*. It is easy to see that this vector field admits a potential* $\varphi(x, y) = xy$*. Let curves* Γ^1 *and* Γ^2 *be given by* $\gamma_1^1(t) = t$, $\gamma_2^1(t) = t$ *and* $\gamma_1^2(t) = t$, $\gamma_2^2(t) = t^2$ *respectively. Note that both curves connect the same points* $(0, 0)$ *and* $(1, 1)$*. According to formula (1.22)*

$$\int_{\Gamma^1} (\mathbf{a}, \mathbf{dx}) = \int_{\Gamma^2} (\mathbf{a}, \mathbf{dx}) = \varphi(1, 1) - \varphi(0, 0) = 1. \tag{1.23}$$

Let us evaluate the curvilinear integrals of $\mathbf{a}$ *along both of these curves directly.*

$$\int_{\Gamma^1} (\mathbf{a}, \mathbf{dx}) = \int_0^1 (t + t)dt = 1 \tag{1.24}$$

$$\int_{\Gamma^2} (\mathbf{a}, \mathbf{dx}) = \int_0^1 (t^2 + 2t^2)dt = 1. \tag{1.25}$$

Therefore, the values of curvilinear integrals along these paths are the same and coincide with (1.23).

1.4 Multiple and Repeated Integrals

Consider a compact set $\Omega \subset R^2$. One can define the integral over Ω in the same way as we define a definite Riemann integral of a function of one variable, if we replace partition of the segment on small segments by a partition of Ω on N^2 small rectangles with sides Δx and Δy and consider the limit of the following expression:

$$\lim_{\Delta x \to 0, \Delta y \to 0} \sum_{i,j=1}^{N} f(\zeta_i, \eta_j) \Delta x \Delta y, \tag{1.26}$$

where (ζ_i, η_j) is some point within rectangle (i, j). This limit, if it exists, is called the double integral of function $f(\cdot)$ over set Ω and is denoted:

$$\iint_{\Omega} h(x, y) dx dy \equiv \lim_{\Delta x \to 0, \Delta y \to 0} \sum_{i,j=1}^{N} f(\zeta_i, \eta_j) \Delta x \Delta y. \tag{1.27}$$

A basic result in the multivariate calculus (the Fubini's Theorem) states that a double integral can be calculated by a reduction to a multiple integral. Namely, let

$$\Omega = \{(x, y) \in R^2 : a \leq x \leq b,\ f(x) \leq y \leq g(x)\} \tag{1.28}$$

for some continuous f and g. Then

$$\iint_{\Omega} h(x, y) dx dy = \int_{a}^{b} (\int_{f(x)}^{g(x)} h(x, y) dy) dx. \tag{1.29}$$

In a similar way one can introduce a repeated integral of order n and reduce it to a multiple integral.

My next objective is to provide a formula that will allow us to transform the multiple integral under a change of variables of integration. Let $\phi : R^n \to R^n$ be a continuously differentiable mapping. Then its Jacobi matrix J is a matrix with a general element

$$J_{ij} = \frac{\partial \phi_i}{\partial x_j}. \tag{1.30}$$

The determinant of this matrix is denoted by J and called the Jacobian. Consider and integral

$$\int \cdots \int h(x) dx_1 ... dx_n \tag{1.31}$$

and let

$$x = \phi(y). \tag{1.32}$$

Then

$$\int \cdots \int_{\phi(\Omega)} h(x) dx_1 ... dx_n = \int \cdots \int_{\Omega} h(\phi(y)) \left| J \right| dy_1 ... dy_n. \tag{1.33}$$

I am not going to prove (1.33) here. An interested reader is referred to Smirnov (1964). However, it is useful to obtain some geometrical intuition about this formula. Divide the set $\phi(\Omega)$ on N small rectangular parallelepipeds with sides parallel to the coordinate axes and having lengths $\Delta x_1, \Delta x_2, ..., \Delta x_n$. Then its volume is given by:

$$V_n = \Delta x_1 \cdot \Delta x_2 \cdot ... \cdot \Delta x_n. \tag{1.34}$$

Let ξ_n be a point inside the parallelepiped and $\chi = \max(\Delta x_1, \Delta x_2, ..., \Delta x_n)$. Then

$$\int \cdots \int_{\phi(\Omega)} h(x) dx_1 ... dx_n = \lim_{\chi \to 0} \sum_{i=1}^{N} h(\xi_n) V_n. \tag{1.35}$$

Under the change (1.32) the initial parallelepipeds will be transformed into ones which are build on vectors $a^1, ..., a^n$ as its sides, where a^i is given by:

$$a^i = \sum_{j=1}^{N} J_{ij}(\zeta_j) \Delta x_j + o(\chi), \tag{1.36}$$

and $\xi_j = \phi(\zeta_j)$. The volume of the transformed parallelepipeds will be given by

$$W_n = |J| \, V_n. \tag{1.37}$$

(Exercise 6 after this chapter will ask you to verify formula (1.37) in the case, when $n = 3$). Therefore, the sum at the right hand side of (1.35) can be written as

$$\sum_{i=1}^{N} h(\zeta_n) W_n + o(\chi^n) = \sum_{i=1}^{N} h(\zeta_n) \, |J(\zeta_n)| \, V_n + o(\chi^n). \tag{1.38}$$

Taking limit as $\chi \to 0$ one obtains equation (1.33).

Below, I am going write down Jacobians for some commonly used changes of variables. The verification of these results is left as an exercise to the reader.

1. Polar coordinates (r, ϕ) on the plane are defined by

$$\begin{cases} x = r \cos \phi \\ y = r \sin \phi \end{cases}. \tag{1.39}$$

If x and y change in the full range, $r \in [0, \infty)$ and $\phi \in [0, 2\pi)$. For this change of variables $J = r$.

2. Cylindrical coordinates (ρ, u, ϕ) in R^3 are defined by

$$\begin{cases} x = \rho \cos \phi \\ y = \rho \sin \phi \\ u = z \end{cases}. \tag{1.40}$$

If x, y, and z change in the full range, $\rho \in [0, \infty)$, $u \in (-\infty, \infty)$ and $\phi \in [0, 2\pi)$. For this change of variables $J = \rho$.

3. Spherical coordinates

$$\begin{cases} x = r \sin\theta \cos\phi \\ y = r \sin\theta \sin\phi \\ z = r \cos\theta \end{cases} . \tag{1.41}$$

If x, y, and z change in the full range, $r \in [0, \infty)$, $\theta \in [0, \pi)$, and $\phi \in [0, 2\pi)$. For this change of variables $J = r^2 \sin\theta$.

One can also define the multiple Lebesgue integral in the usual way starting with the simple functions and then extending it on the set of measurable functions.

Definition 9 *A function $f : \Omega \to R$ is called simple if the set $f(\Omega)$ is finite.*

Let $f : \Omega \to R$ be a finite measurable function and

$$f(\Omega) = \{c_1, ..., c_N\}. \tag{1.42}$$

Define the Lebesgue integral of f over Ω by

$$I(f) = \iint_{\Omega} f(x, y) dx dy \equiv \sum_{j=1}^{N} c_j \lambda_j, \tag{1.43}$$

where λ_j is the Lebesgue measure of the set

$$\{(x, y) \in \Omega : f(x, y) = c_j\}. \tag{1.44}$$

Note that the Lebesgue measure of set (1.44) is well defined, since function $f(\cdot)$ is assumed to be measurable. It can be shown that for any measurable function f the exists a sequence of simple functions $\{f_n(x)\}_{n=0}^{\infty}$ such that

$$\lim_{n \to \infty} f_n(x) = f(x) \tag{1.45}$$

for almost any x and

$$\lim_{n \to \infty} I(f_n) \tag{1.46}$$

depends only on function f itself and not on the sequence chosen. This limit is called the Lebesgue integral of function f.

Note that any Riemann integrable function is Lebesgue integrable, its Riemann and Lebesgue integrals coincide, and the formulae for the change of variables and the Fubini's Theorem apply to the Lebesgue integral as well. There are strictly more Lebesgue integrable functions than Riemann integrable function and the Lebesgue integral has better analytical properties, for example, it is countably additive, while the Riemann integral is not. However, for the purposes of this book the difference between Riemann and Lebesgue integrals is not of a central importance.

1.5 The Flow of a Vector Field and the Gauss-Ostrogradsky Theorem

In the previous Section I introduced the notion of a vector field and defined an important differential operation on vector fields, called divergence. In this Section, I am going to define a certain integral of a vector field, called its flow and establish the connection between the flow of a vector field and its divergence. This, in turn, will allow me to interpret the divergence of a vector field as the density of its sources. Let us start with the following:

Definition 10 *A continuously differentiable mapping* $\mathbf{r} : (0,1)^{n-1} \to R^n$ *is called a smooth surface.*

Define vector $\boldsymbol{\rho}_k(\mathbf{u}) = (\rho_{1k}(\mathbf{u}), ..., \rho_{nk}(\mathbf{u}))$ by

$$\rho_{ik}(u) = \frac{\partial r_i}{\partial u_k}. \tag{1.47}$$

Definition 11 *A smooth surface* $\mathbf{r} : (0,1)^{n-1} \to R^n$ *is called regular if vectors* $\{\ \boldsymbol{\rho}_k(\mathbf{u})\}_{k=1}^{n-1}$ *are linearly independent for all* $\mathbf{u} \in (0,1)^{n-1}$.

For a regular surface define its outside normal vector $\boldsymbol{\nu}(\mathbf{u})$ as the unique vector of a unit length, which is orthogonal to the linear subspace spanned by vectors $\{\boldsymbol{\rho}_k(\mathbf{u})\}_{k=1}^{n-1}$and

$$\det \begin{matrix} \rho_{11}(\mathbf{u}) & ... & \rho_{1n-1}(\mathbf{u}) & \nu_1(\mathbf{u}) \\ \rho_{21}(\mathbf{u}) & ... & \rho_{2n-1}(\mathbf{u}) & \nu_2(\mathbf{u}) \\ ... & ... & ... & ... \\ \rho_{n1}(\mathbf{u}) & ... & \rho_{nn-1}(\mathbf{u}) & \nu_n(\mathbf{u}) \end{matrix} > 0. \tag{1.48}$$

Note that the orthogonality to the $(n-1)$ dimensional subspace spanned by vectors $\{\boldsymbol{\rho}_k(\mathbf{u})\}_{k=1}^{n-1}$ determines the straight line along which the unit vector $\boldsymbol{\nu}$ is aligned, while the last condition fixes its uniquely. Let $S_{n-1}(\mathbf{u})$ be the $(n-1)-$dimensional Lebesgue measure of the parallelepiped built on vectors $\{\boldsymbol{\rho}_k(\mathbf{u})\}_{k=1}^{n-1}$. Then define the surface integral of the second type in the following way.

Definition 12 *A surface integral of the second type of the vector field* $\mathbf{a}$ *over a smooth surface* Σ*, which is defined by mapping* $\mathbf{r} : (0,1)^{n-1} \to R^n$ *is by definition given by:*

$$\Phi_a(\Sigma) = \int \cdots \int_{\Sigma} \mathbf{a} d\Sigma = \int_0^1 ... \int_0^1 (\mathbf{a} \cdot \mathbf{n}) S_{n-1}(\mathbf{u}) d\mathbf{u}$$

and is known as the flow of the vector field through this surface.

Again, it is possible to define the surface integral of the first type, but since we are not going to use it in the book the term *surface integral* will always refer to the surface integral of the second type. The following result is fundamental in vector calculus.

Theorem 13 *(Gauss-Ostrogradsky) Let $\Omega \subset R^n$ be a connected compact set with a piece-wise smooth boundary $\Sigma = \partial\Omega$ and vector field $\mathbf{a} \in C^1(\Omega)$. Then*

$$\int \cdots \int_{\Omega} div\mathbf{a}(\mathbf{x})dx_1...dx_n = \oint_{\Sigma} \mathbf{a}d\Sigma \tag{1.49}$$

This theorem, also known as the *divergence theorem*, links the flow of a vector field to the volume integral of its divergence. It also provides some insight into intuitive meaning of the divergence. To understand it imagine a stream of water running in a valley. Take a point in the stream and encircle it with a small ball. All water which is brought within this imaginary ball by the stream will be taken out. Therefore, the only way for the water flow through the ball to be different from zero is for the stream to have sources within it. We see from this analogy that $div(\mathbf{a})$ represents the density of the sources of the vector field. Let us prove the theorem for the simplest case when $\Omega = \times_{i=1}^{n}[\alpha_i, \beta_i]$ $(\beta_i > \alpha_i)$ is a rectangular parallelepiped. An interested reader is referred to Smirnov (1964) for the general case.

Proof. Consider

$$\int \cdots \int_{\Omega} \frac{\partial a_i}{\partial x_i} dx_1...dx_n. \tag{1.50}$$

Let $\mathbf{x}_{-i}$ be an $(n-1)$ dimensional vector obtained from $\mathbf{x}$ by omitting coordinate x_i and Ω_{-i} be the product $\times_{j=1}^{\prime n}[\alpha_j, \beta_j]$, where $'$ indicates that the term $[\alpha_i, \beta_i]$ is omitted. Then integral (1.50) can be rewritten as

$$\int_{\Omega_{-i}} \Big(\int_{a_i}^{b_i} \frac{\partial a_i}{\partial x_i} dx_i\Big)\mathbf{dx}_{-i} = \int_{\Omega_{-i}} a_i(\mathbf{x}_{-i}, \beta_i)\mathbf{dx}_{-i} - \int_{\Omega_{-i}} a_i(\mathbf{x}_{-i}, \alpha_i)\mathbf{dx}_{-i}. \tag{1.51}$$

Therefore,

$$\int_{\Omega} div\mathbf{a}dx_1...dx_n = \sum_{i=1}^{n}\Big(\int_{\Omega_{-i}} a_i(\mathbf{x}_{-i}, \beta_i)\mathbf{dx}_{-i} - \int_{\Omega_{-i}} a_i(\mathbf{x}_{-i}, \alpha_i)\mathbf{dx}_{-i}\Big). \tag{1.52}$$

On the other hand note that the surface $\partial\Omega = \Sigma$ consists of $2n$ parallelepipeds of dimension $(n-1)$ given by intersection of Ω with hyperplanes $x_i = \alpha_i$ and $x_i = \beta_i$. The unit normal vector to the part of the surface given by $x_i = \alpha_i$ is $\mathbf{n} = (0, .., -1, ...0)$ where -1 is on the ith place, while the unit normal vector to the part of the surface given by $x_i = \beta_i$ is $(0, .., 1, ...0)$

where 1 is on the ith place. Therefore,

$$\oint_{\Sigma} \mathbf{a}d\Sigma = \sum_{i=1}^{n}\Big(\int_{\Omega_{-i}} a_i(\mathbf{x}_{-i}, \beta_i)\mathbf{dx}_{-i} - \int_{\Omega_{-i}} a_i(\mathbf{x}_{-i}, \alpha_i)\mathbf{dx}_{-i}\Big). \tag{1.53}$$

Comparing (1.52) and (1.53) proves the theorem. ■

To conclude this section, it will be useful to provide the expressions for the divergence, the gradient, and the Laplacian is some commonly used curvilinear coordinate systems. For more details, see Zwillinger (2003).

Example 14 *Let us start with writing the expressions for the main operations of vector calculus in polar coordinates (1.39). The coordinates of vector* $\mathbf{a}$ *in polar coordinates* (a_r, a_ϕ) *are linked to its Cartesian coordinates* (a_x, a_y) *by*

$$\begin{cases} a_x = a_r \cos\phi - a_\phi \sin\phi \\ a_y = a_r \sin\phi + a_\phi \cos\phi \end{cases}. \tag{1.54}$$

The main operations of the vector calculus in the polar coordinates are:

$$\begin{cases} div(a) = \frac{1}{r}\frac{\partial(ra_r)}{\partial r} + \frac{1}{r}\frac{\partial a_\phi}{\partial \phi} \\ \nabla f = (\frac{\partial f}{\partial r}, \frac{1}{r}\frac{\partial f}{\partial \phi}) \\ \Delta f = \frac{1}{r}\frac{\partial}{\partial r}(r\frac{\partial f}{\partial r}) + \frac{1}{r^2}\frac{\partial^2 f}{\partial \phi^2} \end{cases} \tag{1.55}$$

Example 15 *The coordinates of vector* $\mathbf{a}$ *in cylindrical coordinates (1.40)* (a_ρ, a_ϕ, a_u) *are linked to its Cartesian coordinates* (a_x, a_y, a_z) *by*

$$\begin{cases} a_x = a_\rho \cos\phi - a_\phi \sin\phi \\ a_y = a_\rho \sin\phi + a_\phi \cos\phi \\ a_z = a_u \end{cases}. \tag{1.56}$$

The main operations of the vector calculus in the cylindrical coordinates are:

$$\begin{cases} div(\mathbf{a}) = \frac{1}{\rho}\frac{\partial(\rho a_\rho)}{\partial \rho} + \frac{1}{\rho}\frac{\partial a_\phi}{\partial \phi} + \frac{\partial a_u}{\partial u} \\ \nabla f = (\frac{\partial f}{\partial \rho}, \frac{1}{\rho}\frac{\partial f}{\partial \phi}, \frac{\partial f}{\partial u}) \\ \Delta f = \frac{1}{\rho}\frac{\partial}{\partial \rho}(\rho\frac{\partial f}{\partial \rho}) + \frac{1}{\rho^2}\frac{\partial^2 f}{\partial \phi^2} + \frac{\partial^2 f}{\partial u^2} \end{cases} \tag{1.57}$$

Example 16 *The coordinates of vector* $\mathbf{a}$ *in spherical (1.41) coordinates* (a_r, a_θ, a_ϕ) *are linked to its Cartesian coordinates* (a_x, a_y, a_z) *by*

$$\begin{cases} a_x = a_r \sin\theta\cos\phi + a_\theta \cos\theta\cos\phi - a_\phi \sin\phi \\ a_y = a_r \sin\theta\sin\phi + a_\theta \cos\theta\sin\phi + a_\phi \cos\phi \\ a_z = a_r \cos\theta - a_\theta \sin\theta \end{cases}. \tag{1.58}$$

The main operations of the vector calculus in the spherical coordinates are:

$$\begin{cases} div(\mathbf{a}) = \frac{1}{r^2}\frac{\partial}{\partial r}(r^2 a_r) + \frac{1}{r\sin\theta}\frac{\partial}{\partial \theta}(a_\theta \sin\theta) + \frac{1}{r\sin\theta}\frac{\partial a_\phi}{\partial \phi} \\ \nabla f = (\frac{\partial f}{\partial r}, \frac{1}{r}\frac{\partial f}{\partial \theta}, \frac{1}{r\sin\theta}\frac{\partial f}{\partial \phi}) \\ \Delta f = \frac{1}{r^2}\frac{\partial}{\partial r}(r^2\frac{\partial f}{\partial r}) + \frac{1}{r^2\sin\theta}\frac{\partial}{\partial \theta}(\sin\theta\frac{\partial f}{\partial \theta}) + \frac{1}{r^2\sin^2\theta}\frac{\partial^2 f}{\partial \phi^2} \end{cases} \tag{1.59}$$

1.6 The Circulation of a Vector Field and the Green Formula

Consider a smooth curve Γ in R^n. The curve is called closed if its start point and its end point coincide. (Caution: do not confuse notion of a closed curve with that of a closed set. For example, a segment is a closed set but not a closed curve). Formally:

Definition 17 *A smooth curve Γ is called closed if $\gamma(0) = \gamma(1)$, where γ is the mapping that defines Γ.*

Let $\mathbf{a}$ be a vector field and Γ be a closed curve.

Definition 18 *A circulation of vector field $\mathbf{a}$ along curve Γ is defined by*

$$circ(\mathbf{a}) = \oint_{\Gamma} (\mathbf{a}, \mathbf{dx}). \tag{1.60}$$

Let $\Omega \subset R^2$ be an open, convex, bounded set with a smooth boundary. There exists an important connection between the circulation of the vector field along the boundary of Ω and the integral of combination of derivatives of the vector field over Ω known as the *Green formula:*

Theorem 19 *Let $\mathbf{a} = (P(x,y), Q(x,y))$ be a vector field on R^2 and $\Omega \subset R^2$ be an open, convex, bounded set with a smooth boundary. Then*

$$\oint_{\partial\Omega} (\mathbf{a}, \mathbf{dx}) = \iint_{\Omega} (\frac{\partial Q}{\partial x} - \frac{\partial P}{\partial y}) dxdy. \tag{1.61}$$

Corollary 20 *Vector field $\mathbf{a}(\cdot)$ is conservative if and only if for any smooth Γ*

$$\oint_{\Gamma} (\mathbf{a}, \mathbf{dx}) = 0. \tag{1.62}$$

I will not prove the Theorem here, but will provide a proof of the corollary. The interested reader is again referred to Smirnov (1964). Note also that though the Corollary is formulated and proved only for the vector field in R^2 the result holds for the vector fields in R^n for any n.

Proof. First, assume that (1.62) holds for any Γ. Consider any two points A and B in Ω and let Γ_1 and Γ_2 be any two paths connecting them. Let Γ_2^- is obtained from Γ_2 by travelling in the opposite direction. Consider a closed curve $\Gamma = \Gamma_1 \cup \Gamma_2^-$. Then

$$0 = \oint_{\Gamma} (\mathbf{a}, \mathbf{dx}) = \oint_{\Gamma_1} (\mathbf{a}, \mathbf{dx}) - \oint_{\Gamma_2} (\mathbf{a}, \mathbf{dx}). \tag{1.63}$$

Therefore, the integral between any two points does not depend on the path taken and Theorem 7 implies that vector field $\mathbf{a}(\cdot)$ admits a potential, i. e. it is conservative.

Now, assume that vector field $\mathbf{a}(\cdot)$ is conservative. Then the exists scalar field $\varphi(\cdot)$ such that $\mathbf{a} = \nabla\varphi$ and Green formula (1.61) implies that

$$\oint_{\Gamma}(\mathbf{a}, \mathbf{dx}) = \oint_{\Gamma}(\frac{\partial^2\varphi}{\partial x\partial y} - \frac{\partial^2\varphi}{\partial y\partial x})dxdy = 0 \tag{1.64}$$

for any smooth closed path Γ. ■

Example 21 *Let* $\mathbf{a} = (-y, x)$ *and* Γ *be a circumference of the unit circle, i. e.* $\gamma(t) = (x(t), y(t))$, *where*

$$\begin{cases} x(t) = \cos(2\pi t) \\ y(t) = \sin(2\pi t) \end{cases}. \tag{1.65}$$

Then

$$\oint_{\partial\Omega}(\mathbf{a}, \mathbf{dx}) = \int_0^1(-y(t)x'(t) + x(t)y'(t))dt. \tag{1.66}$$

Substituting $x(t)$ *and* $y(t)$ *from equation (1.65) one obtains*

$$\oint_{\partial\Omega}(\mathbf{a}, \mathbf{dx}) = 2\pi\int_0^1(\sin^2(2\pi t) + \cos^2(2\pi t))dt = 2\pi. \tag{1.67}$$

One the other hand

$$\frac{\partial Q}{\partial x} - \frac{\partial P}{\partial y} = 2 \tag{1.68}$$

and

$$\iint_{\Omega}(\frac{\partial Q}{\partial x} - \frac{\partial P}{\partial y})dxdy = 2\iint_{\Omega} dxdy = 2\pi, \tag{1.69}$$

since the integral of unity over the unit circle equals its area, namely π.

1.7 Exercises

In all the exercises below $r = \sqrt{x^2 + y^2 + z^2}$ is the distance from point (x, y, z) to the origin.

1. Calculate the divergence of the following vector fields:
   a). $\mathbf{a}_1 = (x, y, z)$ defined on R^3;
   b). $\mathbf{a}_2 = (x/r^3, y/r^3, z/r^3)$ defined on $R^3/\{0\}$.
   c). $\mathbf{a}_3 = (xy - z^2, yz, z)$, defined on R^3.

2. Calculate the gradient and the Laplacian for the following scalar fields (all fields here are defined on $R^3/\{0\}$):
   a). $f = 1/r$;
   b). $f = (a_1 x + a_2 y + a_3 z)/r^5$
3. Which of the following vector fields are conservative:
   a). $\mathbf{a}_1 = (x, z, y)$ defined on R^3;
   b). $\mathbf{a}_2 = (y + 2z, x, y - z)$ defined on R^3;
   c). $\mathbf{a}_3 = (y, -x)$ defined on R^2;
   d). $\mathbf{a}_4 = (x_1 + x_4, x_2 - x_3, x_3, x_4)$ defined on R^4;
4. Calculate a curvilinear integral of $\mathbf{a} = (x + y, -2y)$ along
   a). $\gamma(t) = (t, t)$
   b). $\gamma(t) = (t, t^3)$.
   Note that in both cases $A = (0, 0)$ and $B = (1, 1)$. Do the values of integrals in (a) and (b) coincide.
5. Verify Jacobians for change to polar (1.39), cylindrical (1.40), and spherical (1.41) coordinates.
6. Consider a parallelepiped build on vectors $a = (a_1, a_2, a_3)$, (b_1, b_2, b_3), and $c = (c_1, c_2, c_3)$. Prove that its volume can be calculated as

$$V = |\det A| , \tag{1.70}$$

where matrix A is given by:

$$A = \begin{matrix} a_1 & b_1 & c_1 \\ a_2 & b_2 & c_2 \\ a_3 & b_3 & c_3 \end{matrix} . \tag{1.71}$$

Use this result to verify formula (1.37).
7. Transform the following integrals:
   a). $\iint\limits_{\Omega} f(\sqrt{x^2 + y^2})dxdy$, to polar coordinates, where $\Omega = \{(x, y) \in R^2 : x^2 + y^2 \leq 1\}$.
   b). $\iiint\limits_{\Omega} f(\sqrt{x^2 + y^2}, z)dxdydz$ to cylindrical coordinates, where $\Omega = R^3$.
   c). $\iiint\limits_{\Omega} f(\sqrt{x^2 + y^2 + z^2})dxdydz$ to cylindrical coordinates, where $\Omega = R^3$.
8. Let Σ be a unit sphere in R^3 and $a = (y,\ z,\ x)$. Calculate $\oint\limits_{\Sigma} a d\Sigma$ directly and using the Gauss-Ostrogradsky Theorem. Compare the results.

1.8 Bibliographic Notes

The material covered in this chapter is rather standard and can be found in any undergraduate text on the subject. A very good reference is Smirnov (1964). The book contains numerous examples and exercises, which illustrate all important theoretical concepts. It also contains almost all the prerequisite material for the understanding of this book. A notable exception is the measure theory. A simple exposition of the measure theory and Lebesgue integral can be found in Rudin (1964) or Kolmogorov and Fomin (1970).

2

Partial Differential Equations

An important step in solving a multidimensional screening model is the analysis of a boundary value problem for the systems of partial differential equations. This chapter provides an exposition of basic results and techniques of the theory of partial differential equations of the first and the second order. Material in the first two sections can be found is an easily accessible book by Courant and Hilbert (1989). Another useful reference with many solved examples is Seddon (1957). Therefore, presentation here is more sketchy. The presentation in the section on group theoretic analysis is more detailed, since a complete textbook treatment is harder to access in this area. A useful reference, however, is Cantwell (2002), which also contains some useful software. Throughout this chapter I will use subscripts to denote partial derivatives. Therefore, u_x, for example, will denote the partial derivative of u with respect to x, u_{xy} the second partial derivative with respect to x and y, etc.

2.1 The First Order Partial Differential Equations

Let $u : R^n \to R$ and $\Phi : R^{2n+1} \to R$ be a continuously differentiable functions. A partial differential equation (PDE) of the first order is an expression:

$$\Phi(u, u_{x_1}, ..., u_{x_n}, x_1, .., x_n) = 0. \tag{2.1}$$

In this section I will describe the general methods for solving some classes of partial differential equation. To start, I will introduce two important notions: *the complete integral* and *the general integral.*

2.1.1 The Complete Integral and the General Integral

Consider the following PDE:

$$u_x + \frac{1}{2}u_y^2 = 0. \tag{2.2}$$

It is straightforward to check that

$$u(x, y; a, b) = -\frac{a^2}{2}x + ay + b \tag{2.3}$$

is a solution of (2.2) for any a and b. Therefore, (2.3) is a two-parametric family of solutions of (2.2). Any such family is called a *complete integral* of (2.2). In general let us give the following definition:

Definition 22 *An n – parametric family $u(x; a_1, ..., a_n)$ of solutions of equation (2.1) is called its complete integral.*

Let us return to our example. Put in (2.3) $b = a$, then we will obtain a particular solution of (2.2), namely

$$u(x, y; a) = -\frac{a^2}{2}x + ay + a. \tag{2.4}$$

Define function $v(x, y)$ by

$$v(x, y) = \max_a u(x, y; a). \tag{2.5}$$

It is straightforward to calculate that

$$v(x, y) = \frac{(y+1)^2}{2x}. \tag{2.6}$$

One can check by a direct substitution that $v(x, y)$ solves equation (2.2). This fact is not a coincidence. Indeed, let $w(\cdot)$ is an arbitrary continuously differentiable function. Put in (2.3) $b = w(a)$ and define

$$v(x, y) = \max_a(-\frac{a^2}{2}x + ay + w(a)). \tag{2.7}$$

Then by the envelope theorem $v_x = -a^2/2$ and $v_y = a$, therefore v satisfies equation (2.2). Note that this argument does not rely on the full strength of the assumption that a is the maximizer of $-a^2x/2 + ay + w(a)$. It only uses the fact that a is determined from

$$-ax + y + w'(a) = 0. \tag{2.8}$$

Therefore, the following formulae will describe a solution implicitly

$$\begin{cases} v(x,y) = -a^2x/2 + ay + w(a) \\ -ax + y + w'(a) = 0. \end{cases} \tag{2.9}$$

Note that the family of solutions (2.9) depends on arbitrary function $w(\cdot)$. Such a family is called a *general integral* of (2.2).

Definition 23 *A family of solutions of equation (2.1), which depends on arbitrary continuously differentiable function is called its general integral.*

If one knows the complete integral, the general integral can be constructed along the lines demonstrated in the example. One might think that all solutions of (2.1) can be obtained by varying the arbitrary function in the general integral. This is, however, not always the case. I will illustrate it by an example. To capture all the situations, where the general integral does not contain the entire set of solutions of the PDE, I will introduce a new notion: *the singular integral.*

2.1.2 The Singular Integral

Consider an equation

$$u^2(4 + u_x^2 + u_y^2) = 1. \tag{2.10}$$

It us straightforward to check that a complete integral of equation (2.10) is given by

$$F(x,y,u;a,b) = (x-a)^2 + (y-b)^2 + u^2 - 1 = 0. \tag{2.11}$$

To find the general integral, put in equation (2.11) $b = w(a)$ and write

$$\begin{cases} (x-a)^2 + (y-b)^2 + u^2 = 1 \\ x - a + (y - w(a))w'(a) = 0 \end{cases}. \tag{2.12}$$

Note, however, that there are two solutions of equation (2.10) that are not contained in the general integral (2.12). They are $u = 1$ and $u = -1$. Such solutions are known as *singular integrals*. Note that the in this case the singular integrals can be obtained by eliminating a and b from the system

$$\begin{cases} F(x,y,u;a,b) = 0 \\ F_a(x,y,u;a,b) = 0 \\ F_b(x,y,u;a,b) = 0 \end{cases}. \tag{2.13}$$

This turns out to be a general rule. Given a complete integral

$$F(u, x_1, ..., x_n; a_1, ..., a_n) = 0 \tag{2.14}$$

the general integral is given implicitly by

$$\begin{cases} F(u, x_1, ..., x_n; a_1, ..., a_{n-1}, w(a_1, ..., a_{n-1})) = 0 \\ F_{a_i} + F_{a_n} w_{a_i} = 0 \end{cases} \text{for } i = 1, ..., n-1, \tag{2.15}$$

while the singular integrals are solutions to

$$\begin{cases} F(u, x_1, ..., x_n; a_1, ..., a_n) = 0 \\ F_{a_i} = 0 \end{cases} \text{for } i = 1, ..., n. \tag{2.16}$$

Now let us return to the example from the previous subsection. The complete integral of equation (2.2) is given by

$$F(u, x, y; a, b) = u + \frac{1}{2}a^2 x - ay - b = 0. \tag{2.17}$$

The singular integrals should satisfy

$$\begin{cases} F = u + \frac{1}{2}a^2 x - ay - b = 0. \\ F_a = ax - y = 0 \\ F_b = -1 = 0 \end{cases}. \tag{2.18}$$

Since system (2.18) is incompatible, equation (2.2) has no singular integrals and all its solutions are contained in its general integral.

2.1.3 The Method of Characteristics

Often equations we will encounter in the screening models will have a special structure, namely they will be quasilinear.

Definition 24 *Consider first order PDE (2.1). If*

$$\Phi(u, u_{x_1}, ..., u_{x_n}, x_1, .., x_n) = \sum_{i=1}^{n} a_i(x, u) u_{x_i} + c(x, u) \tag{2.19}$$

then the PDE is called quasilinear. A quasilinear PDE is called linear if all $a_i(x, u)$ *do not depend on* u*, while* $c(x, u)$ *is linear in* u*.*

One can associate with a quasilinear PDE (2.1), (2.19) the following autonomous system of ordinary differential equations

$$\begin{cases} dx_i/d\tau = a_i(x, u) \\ du/d\tau = c(x, u) \end{cases}. \tag{2.20}$$

System (2.20) is known as the system of characteristics.

Definition 25 *Let* $v_i : \Omega \to R$ *be differentiable functions for* $i = 1, .., m$*. These functions are called functionally independent if there exists no differentiable function* $F : R^m \to R$ *not identically equal to zero such that*

$$F(v_1(x), ..., v_m(x)) \equiv 0. \tag{2.21}$$

Note that system (2.20) is a system of $(n+1)$ ordinary differential equations. It is a well-known fact that any such system has n functionally independent first integrals, that is there exist n functionally independent functions $\phi_1(x,u),....,\phi_{n+1}(x,u)$ such that

$$\frac{d\phi_j(x,u)}{d\tau} = \sum_{i=1}^{n} a_i(x,u)\frac{\partial\phi_j(x,u)}{\partial x_i} + c(x,u)\frac{\partial\phi_j(x,u)}{\partial u} = 0. \tag{2.22}$$

Theorem 26 *Let $\phi_1(x,u),....,\phi_n(x,u)$ be functionally independent solutions of (2.20) and F be an arbitrary continuously differentiable function. Then*

$$F(\phi_1(x,u),....,\phi_n(x,u)) \tag{2.23}$$

is a solution of (2.1), (2.19). Moreover, any solution of (2.1), (2.19) has a form (2.23) for some continuously differentiable function F.

Proof. First note that

$$c(x,u)\frac{\partial F}{\partial u} = c(x,u)\sum_{j=1}^{n}\frac{\partial F}{\partial\phi_j}\frac{\partial\phi j(x,u)}{\partial u}. \tag{2.24}$$

From equation (2.22)

$$c(x,u)\frac{\partial\phi j(x,u)}{\partial u} = -\sum_{i=1}^{n} a_i(x,u)\frac{\partial\phi_j(x,u)}{\partial x_i}. \tag{2.25}$$

Therefore,

$$c(x,u)\frac{\partial F}{\partial u} = -\sum_{j=1}^{n}\frac{\partial F}{\partial\phi_j}\sum_{i=1}^{n} a_i(x,u)\frac{\partial\phi_j(x,u)}{\partial x_i}. \tag{2.26}$$

Interchanging the order of summation

$$c(x,u)\frac{\partial F}{\partial u} = -\sum_{i=1}^{n} a_i(x,u)\sum_{j=1}^{n}\frac{\partial F}{\partial\phi_j}\frac{\partial\phi_j(x,u)}{\partial x_i} = -\sum_{i=1}^{n} a_i(x,u)\frac{\partial F}{\partial x_i} \tag{2.27}$$

and $F(\phi_1(x,u),....,\phi_{n+1}(x,u))$ satisfies (2.1), (2.19). The proof of the second part is more involved. An interested reader is referred to Courant and Hilbert (1989). ■

Function F in (2.23) can be determined if one is given the value of the solution on a non-characteristic surface. I will illustrate it below on some examples.

2.1.4 Compatible Systems of the First Order PDEs

Deciding whether a particular systems of the first order partial differential equations is compatible is an important step on the way to deciding implementability of a particular allocation. Therefore, we have to develop a technique that will allow us to address this issue. Let us start with some examples. Consider the following four systems of the first order partial differential equations:

$$\begin{cases} u_x = y \\ u_y = x \end{cases} \tag{2.28}$$

$$\begin{cases} u_x = y + u^2 \\ u_y = u \end{cases} \tag{2.29}$$

$$\begin{cases} u_x = u(x+y) \\ u_y = u(2x+y) \end{cases} \tag{2.30}$$

$$\begin{cases} u_x = y \\ u_y = 2x \end{cases} . \tag{2.31}$$

Our goal is to identify which of these systems are compatible and find the general solutions of the compatible systems.

Note that if $u(x, y)$ solves system

$$\begin{cases} u_x = b_1(x, y, u) \\ u_y = b_2(x, y, u) \end{cases} \tag{2.32}$$

then Young's Theorem (see, Smirnov, 1964) implies that

$$u_{xy} = b_{1y} + b_{1u}b_2 = b_{2x} + b_{2u}b_1 = u_{yx}. \tag{2.33}$$

This condition should hold for all x and y, but in general it need *not* be an identity with respect to u.

Let us start our analysis with system (2.28). In this case (2.33) holds identically. Integrating the first equation of the system one obtains

$$u(x, y) = yx + g(y), \tag{2.34}$$

where $g(\cdot)$ is an arbitrary continuously differentiable function. Substituting it into the second equation one obtains

$$g'(y) = 0. \tag{2.35}$$

Therefore, the general solution to the system (2.28) is

$$u(x, y) = xy + C, \tag{2.36}$$

where C is an arbitrary constant. We see that the solution to the system (2.28) form a uniparametric family. This result is quite general. If condition

(2.33) holds identically the solutions to system (2.32) form a uniparametric family, which can be found by iteratively integrating the system and treating y as a parameter while integrating the first equation and x as a parameter, while integrating the second. The method, clearly generalizes for more than two independent variables.

Now let us move to system (2.29). In that case condition (2.33) implies that

$$u^2 = y - 1. \tag{2.37}$$

Substituting it into system (2.29) one obtains

$$\begin{cases} 0 = 2y - 1 \\ \frac{1}{2\sqrt{y-1}} = \sqrt{y-1} \end{cases}. \tag{2.38}$$

Therefore (2.37) does not satisfy system (2.29) and the latter system is incompatible.

For system (2.30) condition (2.33) implies that

$$u = 0, \tag{2.39}$$

which satisfies system (2.30). Therefore, (2.30) has a unique solution $u = 0$.

Finally, for system (2.31) condition (2.33) implies that

$$1 = 2, \tag{2.40}$$

which is a contradiction. Therefore, system (2.31) is incompatible.

The above examples illustrate that condition (2.33) may be satisfied identically, in which case system (2.32) will have a uniparametric family of solutions, fail to hold for any u, in which case case system (2.32) will be incompatible, or hold for a discrete set U of functions $u(\cdot) : R^2 \to R$, in which case the set of the solutions of (2.32) is a (possibly empty) subset of U. It can be proven that if $b_i(x, y, \cdot)$ is analytic this list of possibilities is exhaustive. The technique can be easily generalized for more than two independent variables.

So far I assumed that the system of the first order partial differential equations is explicitly resolved with respect to the partial derivatives. In principle, one might consider a more general system

$$\begin{cases} f(x, y, u, u_x, u_y) = 0 \\ g(x, y, u, u_x, u_y) = 0 \end{cases}. \tag{2.41}$$

Let

$$J(v, w; x, y) = \det \begin{matrix} v_x & w_x \\ v_y & w_y \end{matrix} \tag{2.42}$$

be the Jacobian of system of functions v, w. Let us introduce new variables p and q by

$$\begin{cases} p = u_x \\ q = u_y \end{cases} \tag{2.43}$$

and assume that

$$J(f,g;p,q) \neq 0. \tag{2.44}$$

Then condition (2.33) should be replaced by

$$J(f,g;x,p) + pJ(f,g;u,p) + J(f,g;y,q) + qJ(f,g;u,q) = 0. \tag{2.45}$$

Again, (2.45) can hold identically, lead to a contradiction or, together with (2.41), determine implicitly three functions $u(x,y)$, $p(x,y)$, and $q(x,y)$. If the latter is the case, one has to check conditions (2.43) to verify whether system (2.41) is compatible. Note also that some (singular) solutions may satisfy the following system

$$\begin{cases} f(x,y,u,p,q) = 0 \\ g(x,y,u,p,q) = 0 \\ J(f,g;p,q) = 0 \\ u_x = p \\ u_y = q \end{cases}. \tag{2.46}$$

2.1.5 *The Method of Characteristics for a Non-quasilinear First Order PDE*

The method of characteristics can be extended from the quasilinear case to cover more general first order partial differential equations. For simplicity, I will illustrate it here only for the case $n = 2$. Generalization for the higher dimensions is straightforward.

Let $f : R^5 \to R$ be a continuously differentiable function and consider a PDE

$$f(x,y,u,u_x,u_y) = 0. \tag{2.47}$$

Define variables p and q by (2.43) and rewrite (2.47) as

$$f(x,y,u,p,q) = 0. \tag{2.48}$$

The idea is to find another function $g : R^6 \to R$ such that system

$$\begin{cases} f(x,y,u,p,q) = 0 \\ g(x,y,u,p,q,a) = 0 \end{cases} \tag{2.49}$$

can be solved for p and q, and for the corresponding solutions $p(x,y,u,a)$ and $q(x,y,u,a)$ the equation

$$du = p(x,y,u,a)dx + q(x,y,u,a)dy \tag{2.50}$$

is integrable, where $a \in R$ is an arbitrary constant. More precisely, I require that the following condition should hold identically[1]:

$$p_y + p_u q = q_x + q_u p. \tag{2.51}$$

---

[1] In principle, (2.51) can hold for a particular $u(x,y,a)$, which in turn satisfies (2.50).

This guarantees that equation (2.50) has a uniparametric family of solutions for any value of constant a, i. e. its general solution has a form

$$F(x,y,u,a,b)=0. \tag{2.52}$$

Expression (2.52) provides us with a complete integral of (2.47). Using the methods developed in Section 2.1.1 one can now find the set of all solutions of (2.47).

The main problem is to find a suitable function g. Note, however, that system (2.49) should be compatible. Therefore, (2.45) implies that it should solve

$$f_p\frac{\partial g}{\partial x}+f_q\frac{\partial g}{\partial y}+(pf_p+qf_q)\frac{\partial g}{\partial u}-(f_x+pf_u)\frac{\partial g}{\partial p}-(f_y+qf_u)\frac{\partial g}{\partial q}=0. \tag{2.53}$$

But this is a linear partial differential equation for g of a type we considered in Section 2.1.3. Its system of characteristics is

$$\frac{dx}{f_p}=\frac{dy}{f_q}=\frac{du}{pf_p+qf_q}=-\frac{dp}{f_x+pf_u}=-\frac{dq}{f_y+qf_u}. \tag{2.54}$$

2.1.6 Examples

Let us consider some examples of application of the method of characteristics.

Example 27 *Let* $\mathbf{a}=(x,y,z)$ *be a vector field on* R^3*. Let us find the general solution of the continuity equation:*

$$u_t+div(u\mathbf{a})=0. \tag{2.55}$$

First rewrite the equation (2.55) in the form

$$u_t+xu_x+yu_y+zu_z=-3u. \tag{2.56}$$

The system of characteristics is

$$\begin{cases} dt/d\tau=1\\ dx/d\tau=x\\ dy/d\tau=y\\ dz/d\tau=z\\ du/d\tau=-3u \end{cases}. \tag{2.57}$$

It is straightforward to check by explicit differentiation that the following functions are the functionally independent first integrals of system (2.57):

$$\phi_1=ye^{-t},\ \phi_2=ze^{-t},\ \phi_3=ue^{3t},\ \phi_4=xe^{-t}. \tag{2.58}$$

The general solution is given by

$$F(xe^{-t}, ye^{-t}, ze^{-t}, ue^{3t}) = 0 \tag{2.59}$$

for arbitrary continuously differentiable function $F : R^4 \to R$. *One can solve this expression explicitly for* u, *to obtain*

$$u = e^{-3t} G(xe^{-t}, ye^{-t}, ze^{-t}) \tag{2.60}$$

for arbitrary continuously differentiable function $G : R^3 \to R$.

Example 28 *Let us find a solution of*

$$u_x + uu_y = 0, \tag{2.61}$$

satisfying the initial condition $u(0, y) = y$. *The system of characteristics has a form*

$$\begin{cases} dx/d\tau = 1 \\ dy/d\tau = u \\ du/d\tau = 0 \end{cases}. \tag{2.62}$$

Two independent first integrals of system (2.62) are

$$\phi_1 = u, \ \phi_2 = y - ux. \tag{2.63}$$

The general solution can be written as

$$u = \eta(y - ux), \tag{2.64}$$

where $\eta(\cdot)$ *is arbitrary continuously differentiable function. From the initial condition we find* $\eta(y) = y$, *therefore*

$$u = y - ux, \tag{2.65}$$

and solving for u

$$u = \frac{y}{1 + x}. \tag{2.66}$$

Example 29 *Let us find the complete integral of the equation*

$$f(u_x, u_y) = 0. \tag{2.67}$$

The uniparametric family of solutions to equation (2.53) is

$$g(x, y, u, p, q, a) = p - a. \tag{2.68}$$

Solving the system

$$\begin{cases} f(p, q) = 0 \\ p - a = 0 \end{cases} \tag{2.69}$$

one obtains

$$\begin{cases} u_x = p = a \\ u_y = q = Q(a) \end{cases}. \tag{2.70}$$

Integrating the above system one find a complete integral

$$u = ax + Q(a)y + b. \tag{2.71}$$

Note that if, on the other hand, one defines $u(\cdot)$ *as*

$$u(x,y) = \max_a (ax + Q(a)y + t(a)), \tag{2.72}$$

where $t(\cdot)$ *is an arbitrary differentiable function, then* $u(\cdot)$ *will solve (2.67) for*

$$f(p,q) = p - Q(q). \tag{2.73}$$

Example 30 *Let us solve the following Cauchy problem:*

$$u_x u_y = 1 \tag{2.74}$$

$$u(0,y) = 2\sqrt{y}. \tag{2.75}$$

Following the steps outlined in the previous example one can obtain a complete integral in the form:

$$u = ax + \frac{y}{a} + b. \tag{2.76}$$

It is easy to check that there are no singular integrals and the general integral is given implicitly by

$$\begin{cases} u = ax + \frac{y}{a} + b(a) \\ x + b'(a) = \frac{y}{a^2} \end{cases}, \tag{2.77}$$

where b *is an arbitrary differentiable function. Using the initial condition*

$$\begin{cases} 2\sqrt{y} = \frac{y}{a} + b(a) \\ b'(a) = \frac{y}{a^2} \end{cases}, \tag{2.78}$$

therefore excluding y *one obtains*

$$(b' - 2 + \frac{b}{a})^2 = 4(1 - \frac{b}{a}). \tag{2.79}$$

It is easy to verify that $b = a$ *solves (2.79) (one can peak up any solution of (2.79) except for* $b = 0$*, since* y *should vary with* a*). Then system (2.77) implies that*

$$a = \sqrt{\frac{y}{x+1}} \tag{2.80}$$

and finally

$$u = 2\sqrt{y(x+1)}. \tag{2.81}$$

2.2 The Second Order Partial Differential Equations

2.2.1 Linear Second Order Partial Differential Equations

A partial differential equation of a form

$$\sum_{i,j=1}^{n} a_{ij}(x)u_{x_i x_j} + \sum_{i=1}^{n} b_i(x)u_{x_i} + c(x)u = f(x) \tag{2.82}$$

is called a linear second order partial differential equation. We will assume that $x \in \Omega$, where $\Omega \subset R^n$ is an open, convex set with a smooth boundary, $\partial\Omega$. We will demand that u satisfy equation (2.82) at Ω, while the behavior on the boundary will be governed by boundary conditions that we will discuss below. The properties of the solution are strongly influenced by the behavior of the quadratic for

$$Q(z;x) = \sum_{i,j=1}^{n} a_{ij}(x)z_i z_j. \tag{2.83}$$

The following main classes of the linear second order PDEs are distinguished in the literature:

Definition 31 *A linear second order PDE is called:*

1. elliptic, if all eigenvalues of the quadratic form $Q(z;x)$ are nonzero and have the same sign for all $x \in \Omega$;

2. parabolic, if all eigenvalues but one of the quadratic form $Q(z;x)$ are nonzero and have the same sign for all $x \in \Omega$, while the remaining one is identically zero.

3. hyperbolic, if all eigenvalues but one of the quadratic form $Q(z;x)$ are nonzero and at least two of them have the different signs for all $x \in \Omega$.

This classification is obviously not exhaustive. However, most interesting equations of mathematical economics and physics fall into one of these types. For example, if $\Omega \subset R^2$, then equation

$$\Delta u = 0, \tag{2.84}$$

called the Laplace equation, is of elliptic type, equation

$$u_x = u_{yy}, \tag{2.85}$$

called the diffusion equation, is of parabolic type, and equation

$$u_{xx} = u_{yy}, \tag{2.86}$$

called the wave equation is of hyperbolic type. All second order PDEs we will meet in the screening applications are of the elliptic type. Parabolic equations found their applications in some other areas of economics: for example, in the theory of financial derivatives pricing and in the theory of bounded rationality. I am not aware of any economic applications of the hyperbolic equations. Below I will take a closer look on elliptic second order PDEs.

2.2.2 Boundary Value Problems for Elliptic Equations

Consider an equation

$$u_{xx} + u_{yy} = 0 \tag{2.87}$$

on set

$$\Omega = \{(x, y) : x^2 + y^2 \leq 1\}. \tag{2.88}$$

This equation has many solutions. For example,

$$u_1(x, y) = e^y \sin x \tag{2.89}$$

and

$$u_2(x, y) = e^x \sin y \tag{2.90}$$

both satisfy (2.87). To select a particular solution one has to impose some conditions that hold on the boundary. The boundary conditions of the first type specify the value of function $u(\cdot)$ on the boundary of set Ω. They lead to the so-called Dirichlet problem:

$$\begin{cases} \sum_{i,j=1}^{n} a_{ij}(x) u_{x_i x_j} + \sum_{i=1}^{n} b_i(x) u_{x_i} + c(x) u = f(x) \\ u(x) = u_0(x) \text{ on } \partial\Omega, \end{cases} \tag{2.91}$$

for some continuous function $u_0(x)$.

Definition 32 *Let functions $a_{ij}(\cdot)$, $b_i(\cdot)$, $c(\cdot)$, and $f(\cdot)$ be twice differentiable on Ω. Then function $u(\cdot)$ twice continuously differentiable on Ω and continuous on its closure, which solves the Dirichlet problem (2.91) is called its classical solution.*

It can be shown that under the assumptions of the definition the classical solutions exists and is unique. In some applications it is not appropriate to assume that the coefficients of the equation (2.91) are well-behaved. In fact, we may be willing to accommodate situations when they are not proper functions at all (e. g. they may be densities of probability distributions with atoms). In that case the concept of the solution for a partial differential equation should be extended. I will discuss exactly how in the last chapter of this part.

The boundary conditions of the second type put some restrictions on the first derivatives of $u(\cdot)$ at the boundary of set Ω. Let us assume that set Ω

has a smooth boundary and $\nu(x)$ denote a unit vector normal to $\partial\Omega$. The Neumann boundary problem is formulated as follows:

$$\begin{cases} div(\mathbf{a}(\mathbf{x}, \nabla u)) = f(\mathbf{x}) \\ \mathbf{a} \cdot \boldsymbol{\nu}(\mathbf{x}) = \xi(\mathbf{x}) \text{ on } \partial\Omega, \end{cases} \tag{2.92}$$

where $\mathbf{a}(\cdot, \cdot)$ is some field and $f(\cdot)$ is a scalar field.

Definition 33 *The function $u(\cdot)$ twice continuously differentiable on Ω and continuous on its closure, which solves the Neumann problem (2.92) is called its classical solution.*

Integrating the first equation in the system (2.92), applying the Gauss-Ostrogradsky theorem (see Theorem 14 in Chapter 1), and taking into account the second equation one obtains

$$\int_{\Omega} \cdots \int f(x)\mathbf{dx} = \int_{\partial\Omega} \cdots \int \xi(\mathbf{x})\mathbf{dx}. \tag{2.93}$$

Therefore, condition (2.93) is necessary for system (2.92) to have a solution. It can be shown that it is also sufficient. More precisely, the following proposition is known to hold:

Proposition 34 *The problem (2.92) has a solution if and only if (2.93) is satisfied, in which case it is unique up to an additive constant.*

One can also consider boundary conditions of the third type, which put some joint restrictions on the function $u(\cdot)$ and its first derivatives on $\partial\Omega$. Such problems arise in some models of boundedly rational behavior. However, since this type of the boundary problems did not yet found its application in screening models, we will not discuss it in the book.

2.2.3 Examples

In this section I solved explicitly some boundary problems. The examples are used not only to illustrate the theoretical points made in the main body of this chapter, but also to familiarize the reader with the practical methods of solving such problems. In doing so I also introduce without a proof some theoretical facts, necessary to justify the steps taken when constructing a solution.

Example 35 *Let Ω be an open set with a piece-wise smooth boundary and consider the following problem*

$$div\mathbf{a} = f(\mathbf{x}) \text{ on } \Omega \tag{2.94}$$

$$\mathbf{a} \cdot \mathbf{n} = 0 \text{ on } \partial\Omega, \tag{2.95}$$

where $f(\cdot)$ is a scalar field. Let us prove that there exists vector field $\mathbf{a}$ *solving the above system if and only if*

$$\int_{\Omega} f(\mathbf{x})\mathbf{dx} = 0. \tag{2.96}$$

The Gauss-Ostrogradsky Theorem (Theorem 14, Chapter 1) immediately implies that (2.96) is necessary for the existence. To prove sufficiency let us look for $\mathbf{a}$ *in a form*

$$\mathbf{a} = \boldsymbol{\nabla}\phi \tag{2.97}$$

for some scalar field $\phi(\cdot)$. Then $\phi(\cdot)$ should solve (2.92) and we know from Proposition 35 that (2.96) is sufficient for the existence of a solution.

The above example will play a particular important role in the development of the Hamiltonian approach to multidimensional screening problems.

Example 36 *Consider the following Dirichlet boundary problem. Let $\Omega = \{(x,y) : x^2 + y^2 < 1\}$. Find $u(x,y)$ such that*

$$\Delta u = 0 \text{ on } \Omega, \tag{2.98}$$

$$u(x,y) = u_0(x,y) \text{ on } \partial\Omega. \tag{2.99}$$

Since equation (2.98) has rotational symmetry (in the next section I will discuss in some detail the question of calculating the symmetries of equations and boundary problems) it is natural to write the problem in the polar coordinates (1.39). Then using equation (1.55) the problem (2.98)-(2.99) can be written as

$$\frac{1}{r}\frac{\partial}{\partial r}(r\frac{\partial u}{\partial r}) + \frac{1}{r^2}\frac{\partial^2 u}{\partial \phi^2} = 0 \text{ for } r < 1 \tag{2.100}$$

$$u(1,\phi) = \Psi(\phi), \tag{2.101}$$

where $\Psi(\phi) = u_0(\cos\phi, \sin\phi)$. Let us look for a solution of problem (2.100)-(2.101) in a form

$$u(r,\phi) = R(r)\Phi(\phi). \tag{2.102}$$

Substituting (2.102) into (2.100) one obtains:

$$\frac{r}{R(r)}\frac{\partial}{\partial r}(rR'(r)) = -\frac{\Phi''(\phi)}{\Phi(\phi)}. \tag{2.103}$$

Since the right hand side of the equation (2.103) depends only on ϕ and the left hand side only on r, both of them should equal some constant, λ. Therefore, one obtains

$$\Phi'' + \lambda\Phi = 0. \tag{2.104}$$

The general solution to equation (2.104) is given by

$$\Phi(\phi) = A\cos(\sqrt{\lambda}\phi) + B\sin(\sqrt{\lambda}\phi). \tag{2.105}$$

Since coordinates (r,ϕ) and $(r,\phi+2\pi)$ refer to the same point of Ω, function $\Phi(\cdot)$ should be $2\pi-$periodic, therefore

$$\sqrt{\lambda} = n \in Z. \tag{2.106}$$

Since constants A and B are arbitrary, one can without loss of generality restrict the attention to the case $n \in N$. Equation for $R(\cdot)$ can now be written in a form

$$R'' + \frac{R'}{r} + n^2\frac{R(r)}{r^2} = 0. \tag{2.107}$$

The general solution of this equation is given by

$$R(r) = \begin{cases} C_{1n}r^n + C_{2n}r^{-n}, & \text{if } n \geq 1 \\ D_1 + D_2 \ln r, & \text{if } n = 0. \end{cases} \tag{2.108}$$

Since the solution should be defined for $r = 0$

$$C_{2n} = D_2 = 0. \tag{2.109}$$

Note that since equation (2.100) is linear any linear combination of the solutions is a solution. Moreover, it can be shown that any solution of (2.100) has this form. This follows from the general fact that the eigenfunctions of a linear elliptic differential operator , i. e. functions that satisfy

$$\sum_{i,j=1}^{n} a_{ij}(x)u_{x_ix_j} + \sum_{i=1}^{n} b_i(x)u_{x_i} + c(x)u = \lambda u \tag{2.110}$$

for some constant λ, form a complete system in the space of twice differentiable functions. See, Courant and Hilbert (1989) for the details. Therefore,

$$u(r,\phi) = D_1 + \sum_{n=1}^{\infty} r^n(A_n\cos(n\phi) + B_n\sin(n\phi)). \tag{2.111}$$

Coefficients D_1, A_n, B_n are found from the boundary condition

$$\Psi(\phi) = D_1 + \sum_{n=1}^{\infty}(A_n\cos(n\phi) + B_n\sin(n\phi)). \tag{2.112}$$

Integrating equation (2.112) from 0 to 2π one obtains

$$D_1 = \frac{1}{2\pi}\int_0^{2\pi}\Psi(\phi)d\phi. \tag{2.113}$$

Multiplying it on $\cos(k\phi)$ *and* $\sin(k\phi)$ *respectively and integrating from* 0 *to* 2π *one obtains*

$$A_k = \frac{1}{\pi} \int_0^{2\pi} \Psi(\phi) \cos(k\phi) d\phi, \tag{2.114}$$

$$B_k = \frac{1}{\pi} \int_0^{2\pi} \Psi(\phi) \sin(k\phi) d\phi. \tag{2.115}$$

Example 37 *Consider the following Dirichlet boundary problem. Let* $\Omega = \{(x,y) : x^2 + y^2 < 1\}$. *Find* $u(x,y)$ *such that*

$$\Delta u = 0 \text{ on } \Omega, \tag{2.116}$$

$$u(x,y) = \ln 2 + x^3 \text{ on } \partial\Omega. \tag{2.117}$$

Following the same steps as in the previous example one obtains

$$u(r,\phi) = D_1 + \sum_{n=1}^{\infty} r^n (A_n \cos(n\phi) + B_n \sin(n\phi)). \tag{2.118}$$

The boundary condition now implies that

$$u(1,\phi) = D_1 + \sum_{n=1}^{\infty} (A_n \cos(n\phi) + B_n \sin(n\phi)) = \ln 2 + \cos^3 \phi. \tag{2.119}$$

Using a well-known formula,

$$\cos(3\phi) = 4\cos^3\phi - 3\cos\phi \tag{2.120}$$

one obtains

$$u(1,\phi) = \ln 2 + \frac{3}{4}\cos\phi + \frac{1}{4}\cos(3\phi). \tag{2.121}$$

Therefore, $D_1 = \ln 2$, $A_1 = 3/4$, $A_3 = 1/4$, $A_n = 0$ *for* $n \notin \{1,3\}$, $B_n = 0$. *Finally, we obtain*

$$u(r,\phi) = \ln 2 + \frac{3r}{4}\cos\phi + \frac{r^3}{4}\cos(3\phi) \tag{2.122}$$

or

$$u(x,y) = \ln 2 + \frac{3x}{4} + \frac{x^3}{4} - \frac{3xy^2}{4}. \tag{2.123}$$

Example 38 *Consider the following Dirichlet boundary problem. Let* $\Omega = \{(x,y) : x^2 + y^2 < 1\}$. *Find* $u(x,y)$ *such that*

$$\Delta u = 1 \text{ on } \Omega, \tag{2.124}$$

$$u(x,y) = \frac{1}{2} \text{ on } \partial\Omega. \tag{2.125}$$

First note that equation (2.124) is linear, therefore its general solution equals the sum of a particular solution and the general solution of the corresponding homogeneous equation. It is easy to check by direct substitution that $u(r) = r^2/4$ *solves (2.124). Therefore,*

$$u(r,\phi) = \frac{r^2}{4} + D_1 + \sum_{n=1}^{\infty} r^n (A_n \cos(n\phi) + B_n \sin(n\phi)). \tag{2.126}$$

Using the boundary condition, $D_1 = 1/4$, $A_n = B_n = 0$. *Therefore,*

$$u(r,\phi) = \frac{1+r^2}{4} \tag{2.127}$$

or

$$u(x,y) = \frac{x^2+y^2+1}{4}. \tag{2.128}$$

The problem might be solved easier if one notice that neither the coefficients of equation (2.124) nor the boundary condition (2.125) depend on angle ϕ. *Therefore, it is possible to find a solution that depends only on* r, *moreover we know that it is the unique solution. Therefore, the problem can be rewritten as*

$$(ru'(r))' = r \tag{2.129}$$

$$u(1) = \frac{1}{2}. \tag{2.130}$$

Now, integrating once

$$ru'(r) = \frac{r^2}{2} + C_1, \tag{2.131}$$

dividing by r *and integrating once more*

$$u(r) = \frac{r^2}{4} + C_1 \ln r + C_2. \tag{2.132}$$

Since the solution should be defined everywhere inside the unit circle, including $r = 0$, *we find that* $C_1 = 0$, *and* $u(1) = 1/2$ *implies that* $C_2 = 1/4$.

In the previous example we used symmetry considerations to reduce a boundary value problem for a partial differential equation to one for an ordinary differential equation. We will see later in the book that this technique is quite general and can be fruitfully used in screening problems. In the next example we solve a Neumann boundary problem.

Example 39 *Consider the following Neumann boundary problem. Let* $\Omega = \{(x,y) : x^2 + y^2 < 1\}$. *Find* $u(x,y)$ *such that*

$$\Delta u = x \text{ on } \Omega, \tag{2.133}$$

$$u_x x + u_y y = 0 \text{ on } \partial\Omega. \tag{2.134}$$

First, note that

$$\iint_{\Omega} x dx dy = \int_{-1}^{1} x dx (\int_{-\sqrt{1-x^2}}^{\sqrt{1-x^2}} dy) = 2 \int_{-1}^{1} x \sqrt{1-x^2} dx = 0. \quad (2.135)$$

The last equality follows from the fact that we integrate an odd function in the symmetric limits. Writing the problem in the polar coordinates (1.39) one obtains.

$$\frac{1}{r}\frac{\partial}{\partial r}(r\frac{\partial u}{\partial r}) + \frac{1}{r^2}\frac{\partial^2 u}{\partial \phi^2} = r\cos\phi \text{ for } r < 1 \quad (2.136)$$

$$u_r = 0 \text{ for } r = 1. \quad (2.137)$$

Note that equation (2.136) is linear therefore, its general solution is the sum of a particular solution and the general solution of corresponding uniform equation (2.100). Let us look for a particular solution in a form

$$u^*(r, \phi) = \xi(r)\cos\phi. \quad (2.138)$$

Then $\xi(\cdot)$ solves

$$\xi'' + \frac{1}{r}\xi' - \frac{1}{r^2}\xi = r. \quad (2.139)$$

It is straightforward to check that

$$\xi(r) = \frac{r^3}{8} \quad (2.140)$$

satisfies this equation. Now

$$u(r, \phi) = \frac{r^3}{8}\cos\phi + D_1 + \sum_{n=1}^{\infty} r^n (A_n \cos(n\phi) + B_n \sin(n\phi)). \quad (2.141)$$

One can write

$$u_r(1, \phi) = \frac{3}{8}\cos\phi + \sum_{n=1}^{\infty}(-A_n n \sin(n\phi) + B_n n \cos n\phi)) = 0. \quad (2.142)$$

Therefore, one finds $B_1 = -3/8$, $A_n = 0$, $B_n = 0$, for $n \geq 2$. Therefore,

$$u(r, \phi) = \frac{r^3}{8}\cos\phi - \frac{3}{8}r\cos\phi + D_1, \quad (2.143)$$

for an arbitrary constant D_1, or

$$u(x, y) = \frac{1}{8}(x^2 + y^2)x - \frac{3}{8}x + D_1. \quad (2.144)$$

2.3 Group Theoretic Analysis of Partial Differential Equations

In the previous chapter we developed some general techniques for dealing with quasilinear partial differential equation of the first and the second order. Unfortunately, no such techniques exist for a general non-linear equations. However, some partial differential equations that arise in screening models are nonlinear. For example, as we will see in the second part of this book, if the dimensionality of a consumer type is greater than the number characteristics the partial derivatives of the consumer surplus will be dependent. This dependence can be captured by a system of non-linear partial differential equations.

Though general methods for solution of a system of non-linear partial differential equations do not exist, some times properties of solutions can be analyzed and even explicit solutions can be found if the problem possesses sufficient symmetry. In the previous chapter we have already used symmetry considerations to reduce a partial differential equation to an ordinary one, observing that the solution should depend on x and y only through the distance from the origin. In this chapter we will develop a regular way to make such arguments.

Symmetry arguments become particularly important in the nonlinear case, since in that case they are often the only hope to arrive at the solution. I will start with a definition of one parameter Lie groups, then describe techniques for calculating Lie groups for a system of PDEs, and finally use it to simplify the problem and sometimes even to arrive at an explicit solution. More detailed exposition of the this material can be found in Cantwell (2002). That book also contains a software for finding Lie groups of many particular PDEs.

2.3.1 One Parameter Lie Groups

I start this section with a definition of a group. Before giving a formal definition let us consider some examples.

Example 40 *Consider set* $X = R/\{0\}$ *and function* $m : X \times X \to X$ *defined by* $m(a,b) = ab$. *This operation* m *satisfies following properties:*

1. $m(a, m(b,c)) = m(m(a,b),c)$ *(or, taking into account definition of* m, $a(bc) = (ab)c$ *), for* $\forall a,b,c \in X$

2. $\exists e \in X : m(a,e) = m(e,a) = a$, *for* $\forall a \in X$ *(namely,* $e = 1$ *)*

3. $\forall a \in X\ \exists b \in X : m(a,b) = e$ *(namely,* $b = 1/a$ *).*

4. $\forall a,b \in X\ m(a,b) = m(b,a)$ *(namely,* $ab = ba$*).*

Example 41 *Consider set* $X = \{A : A$ *is* $n \times n$ *matrix and* $\det A \neq 0\}$ *and function* $m : X \times X \to X$ *defined by* $m(A,B) = AB$. *Operation* m *satisfies following properties:*

1. $m(A, m(B, C)) = m(m(A, B), C)$ *(or, taking into account definition of* m*,* $A(BC) = (AB)C$ *), for* $\forall A, B, C \in X$

2. $\exists E \in X : m(A, E) = m(E, A) = A$*, for* $\forall A \in X$ *(namely,* E *is the identity matrix)*

3. $\forall A \in X\ \exists B \in X : m(A, B) = E$ *(namely,* $B = A^{-1}$ *).*

Example 42 *Consider an arbitrary set* Ω *an let* X *be a set of bijections (one to one mappings) from* Ω *into itself (if* Ω *has some structure on it* X *can be a subset of bijections that preserve it For example, if* Ω *is a topological space,* X *can be the set of homeomorphisms of* Ω *into* Ω*). Define* $m : X \times X \to X$ *in a following way: for any* $A, B \in X$ *define* $C = m(A, B)$ *by*

$$C(\omega) = A(B(\omega)), \text{ for } \forall \omega \in \Omega. \tag{2.145}$$

Then

1. $m(A, m(B, C)) = m(m(A, B), C)$ *(or, taking into account definition of* m*,* $A(BC) = (AB)C$ *), for* $\forall A, B, C \in X$

2. $\exists E \in X : m(A, E) = m(E, A) = A$*, for* $\forall A \in X$ *(namely,* E *is the identity mapping)*

3. $\forall A \in X\ \exists B \in X : m(A, B) = E$ *(namely,* $B = A^{-1}$ *).*

In three examples above we considered different sets endowed with a different operations. However, despite the difference in their nature all three examples exhibit a common structure, in all of them we have a set and a binary operation on it, that satisfies certain properties. This motivates the following definition:

Definition 43 *A set* G *together with a binary operation* $m : G \times G \to G$ *is called a group if the following properties hold:*

a). (Associativity) $m(g_1, m(g_2, g_3)) = m(m(g_1, g_2), g_3)$*, for* $\forall g_1, g_2, g_3 \in G$

b). (Identity) $\exists e \in G : m(g, e) = g$*, for* $\forall g \in G$*. Element* e *is called the identity element of the group.*

c). (Inverse) For $\forall g \in G\ \exists \widetilde{g} \in G : m(g, \widetilde{g}) = e$*. Element* $\widetilde{g} \equiv g^{-1}$ *is called the inverse of* g*.*

If also for $\forall g_1, g_2, \in G$

d). (Commutativity) $m(g_1, g_2) = m(g_2, g_1)$ *the group is called commutative or Abelian.*

It is straightforward to prove the identity element is unique and that each element has a unique inverse. Clearly, a set that contains one element, call it e, together with $m : G \times G \to G$ defined by $m(e, e) = e$ is a group. Such a group is called trivial. A group with more than one element is called non-trivial.

Definition 44 *Let* (G, m) *be a group and let* $H \subset G$*. If* (H, m) *is a group on its own write it is called a subgroup of* (G, m)*.*

To check that (H,m) is a subgroup of (G,m) one has to verify that $m(h_1,h_2) \in H$ for any $h_1,h_2 \in H$, $e \in H$ and $h^{-1} \in H$ for $\forall h \in H$. If there is no confusion about operation m one usually refers to group (G,m) simply as group G.

Groups play important role in the symmetry analysis of equations of any nature for the following reason. Consider a set Ω and let

$$f(\omega) = 0 \tag{2.146}$$

be some equation on Ω. For example, Ω can be the set of real numbers and $f(\cdot)$ a polynomial, or Ω be a set of smooth functions and $f(\cdot)$ be a differential operator. We are interested in a set of bijections of the set Ω that leave equation (2.146) unchanged. An important result is that the set of all such transformations forms a group.

Proposition 45 *Let $G = \{h : \Omega \to \Omega,\ h\text{–is a bijection}\}$ and $H = \{h \in G : (f(\omega) = 0, \omega' = h(\omega)) \Longrightarrow f(\omega') = 0\}$ (that is equation (2.146) is invariant with respect to H). Then H is a subgroup of G.*

Proof. Definition of H implies that $H \subset G$. Therefore, there remains to prove that H is a group. First, note that $e \in H$, since $f(\omega) = 0$ and $\omega' = e(\omega) = \omega$ implies $f(\omega') = 0$. Now, let $h \in H$, then $f(\omega') = 0$ for $\omega' = h(\omega)$, that is transformation $h(\cdot)$ left equation (2.146) unchanged. Consider the inverse transformation $h^{-1}(\cdot)$, then $\omega = h^{-1}(\omega')$ and since $f(\omega) = 0$ one obtains $h^{-1} \in H$. Finally, let $h_1, h_2 \in H$. Then $(f(\omega) = 0, \omega' = h_2(\omega)) \Longrightarrow f(\omega') = 0$ and $(f(\omega') = 0, \omega'' = h_1(\omega')) \Longrightarrow f(\omega'') = 0$, therefore $(f(\omega) = 0, \omega'' = h_1(h_2(\omega))) \Longrightarrow f(\omega'') = 0$ and $h = h_1 \cdot h_2 \in H$. ■

So far we did not impose any structure on the set G. In applications, often groups are realized as groups of transformations of some subset of a Euclidean space. In that case one can impose additional smoothness requirements on the elements of the group. Intuitively, Lie group consists of elements which can be represented as values of an analytical function of some set of real variables. In this paper we will be interested only in the so-called one parametric Lie groups.

Definition 46 *Let $\Xi \subset R^m$ be an open set and $\tau \in R$. Assume that function $F : R^m \times R \to \Xi$ is infinitely differentiable in α and analytic in τ. Consider set G of coordinate transformations*

$$g^\tau : \{\alpha = F(\widetilde{\alpha}, \tau)\} \tag{2.147}$$

together with a binary operation m (composition) defined by

$$m(g^{\tau_1}, g^{\tau_2}) : \{\alpha = F(F(\widetilde{\alpha}, \tau_1), \tau_2)\}. \tag{2.148}$$

If (G,m) is a group, i. e. for

$$\forall \tau_1, \tau_2\ \exists \tau : F(\widetilde{\alpha}, \tau) = F(F(\widetilde{\alpha}, \tau_1), \tau_2) \tag{2.149}$$

it is called a uniparametric Lie group.

Parameter τ is usually chosen in such a way that $g^0 = e$. Let us consider the following example of a uniparametric Lie group.

Example 47 *Consider a set X of the following 2×2 matrices*

$$A(\tau) = \begin{matrix} \cos\tau & \sin\tau \\ -\sin\tau & \cos\tau \end{matrix} \tag{2.150}$$

for $\tau \in [0, 2\pi)$. It is easy to show (you will be asked to do it in Exercise 8 after this chapter) that this set is an Abelian group with respect to the operation of matrix multiplication. Geometrically, matrix (2.150) links coordinates of a vector in two coordinate systems with the same origin, whose axes are rotated at angle τ with respect to each other, that is they define a transformation $F : R^2 \times [0, 2\pi) \to R^2$ by

$$F(\tau, \mathbf{x}) = \begin{cases} x_1 \cos\tau + x_2 \sin\tau \\ -x_1 \sin\tau + x_2 \cos\tau \end{cases}. \tag{2.151}$$

Note that since $F(\tau, \mathbf{x})$ is analytical in all arguments the set of transformations (2.151) together with the operation of composition is a Lie group. This group is known as the group of two-dimensional rotations.

In screening applications we will often need to calculate a symmetry group of a system of equations. This usually can be done by calculating the symmetry groups of each equation separately and then taking their intersection. The following result is crucial for justifying this procedure.

Proposition 48 *Let (G, m) be a group and (H_1, m) and (H_2, m) its subgroups. Let $H = H_1 \cap H_2$. Then (H, m) is a subgroup of (G, m). Moreover, it is a subgroup of (H_1, m) and (H_2, m).*

Proof. First, note that since both H_1 and H_2 are subgroups of G the identity element $e \in H_1$ and $e \in H_2$, therefore $e \in H_1 \cap H_2$. Let $h \in H_1 \cap H_2$, then $h \in H_1$ and $h \in H_2$. Again, since both H_1 and H_2 are subgroups of G the inverse element $h^{-1} \in H_1$ and $h^{-1} \in H_2$, therefore $h^{-1} \subset H_1 \cap H_2$. Finally, let $h_1, h_2 \in H_1 \cap H_2$. Then $h_1, h_2 \in H_1$ and $h_1, h_2 \in H_2$. since both H_1 and H_2 are subgroups of G the product $h = h_1 \cdot h_2$ belongs to both H_1 and H_2, therefore $h \in H_1 \cap H_2$ and the conclusion follows. ■

Again, if there is no confusion about the nature of the group multiplication, we will simply phrase the proposition as: An intersection of two subgroups is a subgroup.

2.3.2 Invariance of PDEs, Systems of PDEs, and Boundary Problems under Lie Groups

Let us specialize our analysis of the symmetries of general equations to a particular case when the equation in question a second order partial differential equation

$$\Phi(\alpha, u, \nabla u, D^2 u) = 0, \tag{2.152}$$

where $\Phi : R^{m^2/2+5m/2+1} \to R$ is a twice continuously differentiable function. Consider a transformation of the independent and dependent variables:

$$\begin{cases} \widetilde{\alpha}_i = F_i(\boldsymbol{\alpha}, u; \tau) \\ \widetilde{u} = G(\boldsymbol{\alpha}, u; \tau) \end{cases}, \tag{2.153}$$

where functions F_i and G are infinitely differentiable in α and u and analytic in τ, and $F_i(\alpha, u; 0) = \alpha_i$, $G(\alpha, u; 0) = u$. Assume also that transformations (2.153) form a group.

Definition 49 *Equation (2.152) is invariant under a uniparametric Lie group of transformations (2.153) if (2.152) and (2.153) imply that*

$$\Phi(\widetilde{\alpha}, \widetilde{u}, \nabla\widetilde{u}, D^2\widetilde{u}) = 0. \tag{2.154}$$

For any two groups G_1 and G_2, let us write $G_1 \succeq G_2$ if G_2 is a subgroup of G_1. It is left as an exercise to the reader to establish that $\succeq$ is a partial order.

Definition 50 *A symmetry group of equation (2.153) is any uniparametric Lie group that leaves it invariant. The complete symmetry group of equation (2.153) is the maximal Lie group under partial order $\succeq$ that leaves equation (2.153) invariant.*

Definition 51 *A system of partial differential equation is invariant under a uniparametric Lie group of transformations (2.153) if each its equation is invariant.*

The following result is straightforward and its proof is left as an exercise to the reader.

Proposition 52 *Let $H_1, ..., H_m$ are symmetry groups of equations*

$$\Phi_i(\boldsymbol{\alpha}, u, \nabla u, D^2 u) = 0 \tag{2.155}$$

respectively. Then

$$H = \bigcap_{i=1}^{n} H_i \tag{2.156}$$

is the symmetry group of the system

$$\Phi_i(\boldsymbol{\alpha}, u, \nabla u, D^2 u) = 0, \ i = 1, ..., m. \tag{2.157}$$

Moreover, if $H_1, ..., H_m$ are the complete symmetry groups of the corresponding equations, then H defined by (2.156) is the complete symmetry group of system (2.157).

Finally we have to define a symmetry group of a boundary problem. The new issue that arises is that now not only the equations and the boundary conditions should remain unchanged, but also the domain and its boundary. Alternatively, one may include the transformations of the parameters describing the domain into the analog of system (2.153). If we take the first approach we will arrive at the following definition.

Definition 53 *Let $\Omega \subset R^m$ be an open set with a piece-wise smooth boundary $\partial\Omega$. A boundary problem*

$$\begin{aligned} \Phi(\boldsymbol{\alpha}, u, \nabla u, D^2 u) &= 0, \text{ on } \Omega \qquad (2.158)\\ G(\boldsymbol{\alpha}, u, \nabla u) &= 0, \text{ on } \partial\Omega \qquad (2.159) \end{aligned}$$

is invariant with respect to (2.153) if
1. Under transformations (2.153) Ω maps into Ω and $\partial\Omega$ into $\partial\Omega$
2. Equations (2.158) and (2.159) are invariant under (2.153).

This definition is useful when both the domain and the equations possess some obvious symmetry (e.g. a circle possesses rotational symmetry and all coefficients in the equation and the boundary condition depend on the distance from the center). In many interesting applications, however, these conditions will not be satisfied. For example, in a screening model with a rectangular type space and polynomial utilities, the partial differential equations describing the optimal allocation will be invariant with respect to rescaling the axes, but the boundary conditions will not.

To overcome this problem, let us consider another approach. For the sake of concreteness, assume

$$\Omega = \prod_{i=1}^{n} (0, a_i). \qquad (2.160)$$

Then the boundary problem can be written in a way

$$\begin{aligned} \Phi(\boldsymbol{\alpha}, u, \nabla u, D^2 u) &= 0, \text{ on } \Omega \qquad (2.161)\\ G(\mathbf{a}, \boldsymbol{\alpha}, u, \nabla u) &= 0, \text{ on } \partial\Omega. \qquad (2.162) \end{aligned}$$

Definition 54 *Let $\Omega \subset R^m$ be an open set with a piece-wise smooth boundary $\partial\Omega$. A boundary problem*

$$\begin{aligned} \Phi(\boldsymbol{\alpha}, u, \nabla u, D^2 u) &= 0, \text{ on } \Omega \qquad (2.163)\\ G(\mathbf{a}, \boldsymbol{\alpha}, u, \nabla u) &= 0, \text{ on } \partial\Omega \qquad (2.164) \end{aligned}$$

is invariant with respect to the uniparametric Lie group of transformations

$$\begin{cases} \widetilde{\alpha}_i = F_i(\boldsymbol{\alpha}, u; \tau) \\ \widetilde{u} = G(\boldsymbol{\alpha}, u; \tau) \\ \widetilde{a}_i = F_i(a_i, \boldsymbol{\alpha}_{-i}, \boldsymbol{\tau}) \end{cases}, \qquad (2.165)$$

if both equations (2.163) and (2.164) are invariants with respect to it.

In the above definition we used the convention that $\boldsymbol{\alpha}_{-i}$ denotes all components of vector $\boldsymbol{\alpha}$ except for α_i. The augmented transformation (2.165) imply that the boundary $\widetilde{\alpha_i} = \widetilde{a_i}$ is the image of the boundary $\alpha_i = a_i$. Our next objective is to develop some regular techniques for calculating the symmetry groups for the systems of PDEs and boundary problems.

2.3.3 Calculating a Lie Group of a PDE

Let us start with calculating a Lie group for a single PDE. For this purpose define

$$\begin{cases} \theta^j = \frac{\partial F}{\partial \tau}(\alpha, u; 0) \\ \chi = \frac{\partial G}{\partial \tau}(\alpha, u; 0) \end{cases} . \tag{2.166}$$

Then up to $O(\tau)$ terms:

$$\begin{cases} \widetilde{\alpha_i} = \alpha_i + \tau\theta_i(\alpha, u) \\ \widetilde{u} = u + \tau\chi(\alpha, u) \end{cases} . \tag{2.167}$$

Expression (2.167) is known as the infinitesimal form of (2.153) and functions $\theta_i(\alpha, u)$ and $\chi(\alpha, u)$ are called the *group generators*. To calculate the symmetry group of equation (2.158) one has first to extend (2.167) to cover the transformations of the first and second derivatives of u. In doing so, one arrives at the so-called twice-extended group:

$$\begin{cases} \widetilde{\alpha_i} = \alpha_i + \tau\theta^i(\alpha, u) \\ \widetilde{u} = u + \tau\chi(\alpha, u) \\ \widetilde{u_i} = u_i + \tau\chi_{\{i\}}(\alpha, u, \nabla u) \\ \widetilde{u_{ij}} = u_{ij} + \tau\chi_{\{ij\}}(\alpha, u, \nabla u, D^2 u) \end{cases} , \tag{2.168}$$

where

$$\begin{cases} u_i = \partial u/\partial\alpha_i, u_{ij} = \partial^2 u/\partial\alpha_i\partial\alpha_j \\ \chi_{\{i\}} = D_i\chi - \sum_{j=1}^{m} u_j D_i\theta^j \\ \chi_{\{ij\}} = D_i\chi_{\{j\}} - \sum_{k=1}^{m} u_{jk} D_i\theta^k \end{cases} , \tag{2.169}$$

and the total differentiation operator D_i is defined by

$$D_i\omega(\alpha, u) = \frac{\partial\omega}{\partial\alpha_i} + u_i\frac{\partial\omega}{\partial u}. \tag{2.170}$$

The invariance group of equation (2.158) can then be found from the condition

$$\sum_{j=1}^{m} \theta^j \frac{\partial\Phi}{\partial\alpha_j} + \chi\frac{\partial\Phi}{\partial u} + \sum_{k=1}^{m} \chi_{\{i\}}\frac{\partial\Phi}{\partial u_i} + \sum_{k=1}^{m} \chi_{\{ij\}}\frac{\partial\Phi}{\partial u_{ij}} = 0, \tag{2.171}$$

which should hold on the surface $\Phi(\alpha, u, \nabla u, D^2u) = 0$. Carrying out explicit calculations will result is a system of partial differential equations for functions (θ^i, χ). Since we have to find *a* symmetry group, we will be usually interested in a particular finite parametric set of solutions to the system. Cantwell (2002) contains a software that can deal with the problem.

If one has to find a symmetry group of a system of PDEs

$$\Phi_i(\alpha, u, \nabla u, D^2u) = 0, i = 1, .., p \tag{2.172}$$

the above technique can be used to calculate a symmetry groups H_i of each of the equations. Then

$$H = \bigcap_{i=1}^{n} H_i \tag{2.173}$$

will be the symmetry group of the system.

A boundary problems can be represented as system (2.163)- (2.164). The symmetry transformations should be augmented according to (2.165) and then one can treat the boundary problem as a special kind of a system.

2.3.4 Calculating Invariants of the Lie Group

Consider a uniparametric Lie group of transformations given by (2.153), whose infinitesimal form is given by (2.167). The main idea behind calculating the invariance group is to calculate its infinitesimal form and then to integrate to obtain the finite form. I will not try to justify this approach here. An interested reader should see Cantwell (2002).

Definition 55 *A continuously differentiable function* $\Upsilon : R^{m^2/2+5m/2+1} \to R$ *is called an invariant of group (2.153) if* $\forall \tau > 0$

$$\Upsilon(\widetilde{\alpha}, \widetilde{u}, \widetilde{\nabla u}, \widetilde{D^2u}) = \Upsilon(\alpha, u, \nabla u, D^2u). \tag{2.174}$$

Here $\widetilde{\nabla u}$ and $\widetilde{D^2u}$ are calculated using twice extended group (2.169). Expression on the left hand side of (2.174) can be viewed as a function of the group parameter τ and the invariance condition can be read to say that it does not depend on τ, therefore

$$\frac{d\Upsilon}{d\tau} = 0 \tag{2.175}$$

or, taking the full derivative of (2.174) and using (2.169):

$$\sum_{j=1}^{m} \theta^j \frac{\partial \Upsilon}{\partial \alpha_j} + \chi \frac{\partial \Upsilon}{\partial u} + \sum_{k=1}^{m} \chi_{\{i\}} \frac{\partial \Upsilon}{\partial u_i} + \sum_{k=1}^{m} \chi_{\{ij\}} \frac{\partial \Upsilon}{\partial u_{ij}} = 0. \tag{2.176}$$

If the transformation affects not only the coordinates α but also vectors of parameters a and b, they should be treated as additional arguments in

Υ. Note that (2.176) is a linear homogenous partial differential equation. Often such an equation can be solved explicitly. We see that the problem of finding the invariants of a group is easier than finding the symmetry group. However, while for the last problem we usually are interested in finding *a* solution, for this problem we are usually interested in finding all independent invariants.

Consider an important case when $\chi = 0$ (pure coordinate transformation) and suppose we are interested in finding and invariant a function of α that is invariant with respect to (2.153). Then (2.176) reduces to

$$\sum_{j=1}^{m} \theta^j \frac{\partial \Upsilon}{\partial \alpha_j} = 0. \tag{2.177}$$

This is the case that arises in the screening applications. The role of Υ is played by the consumer surplus function.

2.3.5 *Examples*

Example 56 *Let us calculate the symmetry group of an ordinary differential equation*

$$u_{tt} = 0. \tag{2.178}$$

According to equation (2.171) the group generators should satisfy

$$\chi_{\{tt\}} = D_t \chi_{\{t\}} - u_{tt} D_t \theta^t = 0, \tag{2.179}$$

provided $u_{tt} = 0$. *Using*

$$\chi_{\{t\}} = D_t \chi - u_t D_t \theta, \tag{2.180}$$

and the definition of the total differentiation operator D_t :

$$D_t \omega(\alpha, u) = \frac{\partial \omega}{\partial \alpha_t} + u_t \frac{\partial \omega}{\partial u} \tag{2.181}$$

one obtains (after omitting terms proportional to u_{tt}*)*

$$\chi_{tt} + 2u_t \chi_{tu} + u_t^2 \chi_{uu} - u_t \theta_{tt} - 2u_t^2 \theta_{tu} - u_t^3 \theta_{uu} = 0. \tag{2.182}$$

Since this should hold for any u_t *one obtains*

$$\begin{cases} \chi_{tt} = 0 \\ \chi_{tu} = 0 \\ \chi_{uu} = 0 \\ \theta_{tt} = 0 \\ \theta_{tu} = 0 \\ \theta_{uu} = 0 \end{cases} . \tag{2.183}$$

Therefore,

$$\begin{cases} \theta(u,t) = a_{11}t + a_{12}u + b_1 \\ \chi(u,t) = a_{21}t + a_{22}u + b_2 \end{cases}, \tag{2.184}$$

for some real constants a_{11}, a_{12}, a_{21}, and a_{22}. Writing transformations that leave equation (2.178) invariant in the finite form:

$$\begin{pmatrix} \widetilde{t} \\ \widetilde{u} \end{pmatrix} = A(\tau) \begin{pmatrix} t \\ u \end{pmatrix} + \mathbf{c}(\tau), \tag{2.185}$$

for some non-degenerate matrix $A(\tau)$ and some vector $\mathbf{c}(\tau)$ such that $A(0) = I$, $\mathbf{c}(0) = \mathbf{0}$.

Example 57 *Let us find the general form of a function $u(x,t)$ invariant with respect to*

$$\begin{cases} \widetilde{x} = e^{2\tau}x \\ \widetilde{t} = e^{\tau}t \\ \widetilde{u} = u \end{cases}. \tag{2.186}$$

First, note that

$$\begin{cases} \theta_1 = \partial\widetilde{x}/\partial\tau|_{\tau=0} = 2x \\ \theta_2 = \partial\widetilde{t}/\partial\tau|_{\tau=0} = t \\ \chi = \partial\widetilde{u}/\partial\tau|_{\tau=0} = 0 \end{cases}, \tag{2.187}$$

therefore, any function $u(x,t)$ invariant with respect to (2.186) solves

$$2x\frac{\partial u}{\partial x} + t\frac{\partial u}{\partial t} = 0. \tag{2.188}$$

The system of characteristics has a form

$$\begin{cases} dt/ds = t \\ dx/ds = 2x \\ du/ds = 0 \end{cases}. \tag{2.189}$$

The two independent first integrals of this system are

$$\frac{x}{t^2} = C_1, \;\; u = C_2. \tag{2.190}$$

Therefore, the general solution is

$$u(x,t) = u(\frac{x}{t^2}). \tag{2.191}$$

2.4 Exercises

1. Find a complete integral of the equations:
   a).

$$(u_x^2 + u_y^2)y = u_y u \tag{2.192}$$

b).

$$u_x = (u + yu_y)^2 \tag{2.193}$$

2. Solve a Cauchy problem

$$u_x + u_y^2 + xu_x = 0 \tag{2.194}$$
$$u(1, y) = 1. \tag{2.195}$$

3. Let $\Omega = \{(x, y, z) \in R^3 : x^2 + y^2 + z^2 < 1\}$. Solve the following Dirichlet problems:

a).

$$\Delta u = 1 \tag{2.196}$$
$$u(x, y, z) = 4 \text{ on } \partial\Omega. \tag{2.197}$$

b).

$$\Delta u + u = \ln(1 + x^2 + y^2 + z^2) \tag{2.198}$$
$$u(x, y, z) = 0 \text{ on } \partial\Omega. \tag{2.199}$$

4. Let $\Omega = \{(x, y, z) \in R^3 : x^2 + y^2 < 1,\ |z| < 1\}$. Solve the following Dirichlet problems:

a).

$$\Delta u = x \tag{2.200}$$
$$u(x, y, z) = 0 \text{ on } \partial\Omega. \tag{2.201}$$

b).

$$\Delta u = z \tag{2.202}$$
$$u(x, y, z) = 0 \text{ on } \partial\Omega. \tag{2.203}$$

5. Let $\Omega = \{(x, y) \in R^2 : x^2 + y^2 < 1\}$. Determine whether the following Neumann problems have solution and , if yes, find them

a).

$$\Delta u = y \tag{2.204}$$
$$\nabla u \cdot \nu = x \text{ on } \partial\Omega \tag{2.205}$$

b).

$$\Delta u = y \tag{2.206}$$
$$\nabla u \cdot \nu = x^2 \text{ on } \partial\Omega. \tag{2.207}$$

c).

$$\Delta u = y + ax^2 \tag{2.208}$$
$$\nabla u \cdot \nu = y^2 \text{ on } \partial\Omega, \tag{2.209}$$

where a is some constant.
6. Give examples of three Abelian and three non-commutative groups.
7. Prove that in any group the identity element is unique and each element has a unique inverse.
8. Prove that set of matrices (2.150) for $\tau \in [0, 2\pi)$ is an Abelian group. (Hint: using

$$\cos(\tau_1 + \tau_2) = \cos\tau_1 \cos\tau_2 - \sin\tau_1 \sin\tau_2 \tag{2.210}$$
$$\sin(\tau_1 + \tau_2) = \sin\tau_1 \cos\tau_2 + \cos\tau_1 \sin\tau_2 \tag{2.211}$$

prove that

$$A(\tau_1) \cdot A(\tau_2) = A(\tau_1 + \tau_2), \tag{2.212}$$

and using equation (2.212) find the identity element of the group and the inverse element of an element of the group).
9. Prove that the Laplace equation on R^2 is invariant with respect to the group of rotations.
10. Find an invariance group of the diffusion equation

$$u_t = u_{xx}. \tag{2.213}$$

2.5 Bibliographic Notes

Material of the first two sections of this chapter is rather standard and can be found in numerous textbooks on the subject. The classical exposition in Courant and Hilbert (1989) probably still remains the best and the most complete. Another excellent reference with numerous solved examples is Tikhonov and Samarski (1964). A useful more elementary reference with an abundance of examples is Seddon (1957).

Though the results of Sophus Lie on the application of group theory to the theory of ordinary and partial differential equations are more than hundred years old, there are rather few good expositions of the subject. The best for the beginner who wants to master quickly the practical applications is the book of Cantwell (2002). It contains numerous examples and a disk with a software that allows to find symmetry groups for some standard equations.

It is also useful to note that group theoretic analysis was already used in economics in other contexts. See, for example, Sato and Ramachandran (1990) and the references there.

3

Theory of Generalized Convexity

One of way to define the convex functions is through noting that they are exactly the functions that can be represented as suprema of the affine functions, that is $s(\mathbf{x})$ is convex if and only if

$$s(x) = \sup_{a,b \in \Omega} (\mathbf{a} \cdot \mathbf{x} + b) \tag{3.1}$$

for some $\mathbf{a} \in R^n$ and $b \in R$ and some $\Omega \subset R^{n+1}$. Many properties of the convex functions are the result of their representation (3.1). Theory of generalized convexity studies exactly those properties, replacing the family of the affine functions by some arbitrary family.

Good guides to theory of generalized convexity are Rubinov (2000) and Singer (1997). The theory already found some applications in economics, for example, in the field of non-smooth global optimization. Recently, it was shown that it has an important role to play in multidimensional screening problems, when Carlier (2002) and Basov (2002) demonstrated that a consumer surplus function is implementable by a tariff if and only if it is generalized convex. Moreover, the implementing tariff is the generalized convex conjugate of the surplus. The usefulness of this criterion is that the generalized convexity can be checked in a straightforward way: A function is generalized convex if and only if it is equal to its generalized biconjugate.

In this chapter we define the generalized convex functions and study their properties. We also study the relation between generalized convexity of the function and cyclic monotonicity of its subdifferential. Characterization of implementability in terms of cyclic monotonicity is due to Rochet (1987).

This criterion, however, unlike the generalized convexity is notoriously difficult to check.

3.1 The Generalized Fenchel Conjugates

Suppose $\Omega \subset R^m$ is an open, bounded, convex set and $s : \Omega \to R$ is a function. Let $R^* = R \cup \{-\infty, \infty\}$ be the extended real line. That is the extended real line is a real line supplemented with two ideal elements, the positive and the negative infinity. I will also extend the linear order and operations of addition and subtraction on R^* assuming that for $\forall x \in R$ the following relations hold:

$$\begin{cases} -\infty < x < \infty \\ x + \infty = \infty \\ x - \infty = -\infty \\ -\infty + \infty = \infty \end{cases} . \tag{3.2}$$

Let $\overline{\Omega}$ be the closure of Ω. Given a continuous function $u : \overline{\Omega} \times R^n \to R$, define a generalized Fenchel $u-$conjugate in the following way.

Definition 58 *Function $s^* : R^n \to R^*$ defined by*

$$s^*(\mathbf{x}) = \sup_{\alpha \in \Omega}(u(\boldsymbol{\alpha}, \mathbf{x}) - s(\boldsymbol{\alpha})) \tag{3.3}$$

is called $u-$conjugate of $s(\boldsymbol{\alpha})$.

Note that supremum in the above definition can in principle be infinite. Therefore, $s^*(\cdot)$ maps R^n into the extended real line.

In a similar way one can define generalized Fenchel conjugates for functions from R^n into real line, namely

$$t^*(\boldsymbol{\alpha}) = \sup_{\mathbf{x} \in R^n} (u(\boldsymbol{\alpha}, \mathbf{x}) - t(\mathbf{x})). \tag{3.4}$$

Let us now define a function to be $u-$convex if it can be represented in form (3.4) for some function $t(\cdot)$. More precisely,

Definition 59 *Function $s(\cdot)$ is called $u-$convex if $\exists t : R^n \to R^*$ such that*

$$s(\boldsymbol{\alpha}) = \sup_{\mathbf{x} \in R^n} (u(\boldsymbol{\alpha}, \mathbf{x}) - t(\mathbf{x})). \tag{3.5}$$

Let us start by proving the following lemma.

Lemma 60 *Let $\overline{\mathbf{x}} \in R^n$, and $\overline{t} \in R$ then $s(\cdot)$ defined by*

$$s(\boldsymbol{\alpha}) = u(\boldsymbol{\alpha}, \overline{\mathbf{x}}) + \overline{t} \tag{3.6}$$

is $u-$convex.

Proof. Define $t : R^n \to R^*$ by

$$t(x) = \begin{cases} -\bar{t}, & \text{for } \mathbf{x} = \bar{\mathbf{x}} \\ \infty, & \text{otherwise} \end{cases} . \tag{3.7}$$

Then

$$s(\boldsymbol{\alpha}) = \sup_{\mathbf{x} \in R^n} (u(\boldsymbol{\alpha}, \mathbf{x}) - t(\mathbf{x})). \tag{3.8}$$

■

The above lemma states that if vector $\mathbf{x}$ is fixed the resulting function will be $u-$convex. It generalizes a well-known fact that a linear function is convex.

It is a well known fact in the theory of convex functions that a function is convex if and only if it is equal to the convex conjugate of its convex conjugate, i. e. if and only if it coincides with its *biconjugate*. It turns out that a similar criterion holds in the case of the generalized convex functions.

Definition 61 *Function $s^{**}(\alpha)$ defined by*

$$s^{**}(\boldsymbol{\alpha}) = \sup_{\mathbf{x} \in R^n} (u(\boldsymbol{\alpha}, \mathbf{x}) - s^*(\mathbf{x}))$$

is called $u-$biconjugate of $s(\boldsymbol{\alpha})$.

It is easy to see that for any function $s(\cdot)$:

$$s^{**}(\boldsymbol{\alpha}) \leq s(\boldsymbol{\alpha}). \tag{3.9}$$

The proof of this assertion is left as an exercise to the reader. The following theorem summarizes the main result of this Section.

Theorem 62 *A function $s(\cdot) : \Omega \to R$ is $u-$convex if and only if*

$$s(\boldsymbol{\alpha}) = s^{**}(\boldsymbol{\alpha}) \text{ for } \forall \boldsymbol{\alpha} \in \Omega. \tag{3.10}$$

Proof. Suppose $s(\boldsymbol{\alpha}) = s^{**}(\boldsymbol{\alpha})$ for $\forall \boldsymbol{\alpha} \in \Omega$. Then

$$s(\boldsymbol{\alpha}) = \sup_{\mathbf{x} \in R^n} (u(\boldsymbol{\alpha}, \mathbf{x}) - s^*(\mathbf{x})), \tag{3.11}$$

and therefore is $u-$convex.

Now, suppose that $s(\boldsymbol{\alpha})$ is $u-$convex. Then there exists $t : R^n_+ \to R^*$ such that

$$s(\boldsymbol{\alpha}) = \sup_{\mathbf{x} \in R^n} (u(\boldsymbol{\alpha}, \mathbf{x}) - t(\mathbf{x})). \tag{3.12}$$

Let $X \subset R^n$ be the set of points, where $t(\cdot)$ is finite. (Note that $t(\cdot)$ is never equal to $-\infty$, otherwise $s(\cdot)$ would be identically equal to $+\infty$, which we assumed not to be the case). Then

$$s(\boldsymbol{\alpha}) = \sup_{\mathbf{x} \in X} (u(\boldsymbol{\alpha}, \mathbf{x}) - t(\mathbf{x})). \tag{3.13}$$

and for any $(\alpha,x) \in \Omega \times X$

$$t(\mathbf{x}) \geq u(\boldsymbol{\alpha}, \mathbf{x}) - s(\boldsymbol{\alpha}). \tag{3.14}$$

therefore, taking suprema of the both sides

$$t(\mathbf{x}) \geq s^*(\mathbf{x}). \tag{3.15}$$

Therefore,

$$s(\boldsymbol{\alpha}) = \sup_{\mathbf{x}\in X}(u(\boldsymbol{\alpha}, \mathbf{x}) - t(\mathbf{x})) \leq \sup_{\mathbf{x}\in X}(u(\boldsymbol{\alpha}, \mathbf{x}) - s^*(\mathbf{x})) = s^{**}(\boldsymbol{\alpha}) \leq s(\boldsymbol{\alpha}), \tag{3.16}$$

which implies

$$s(\boldsymbol{\alpha}) = s^{**}(\boldsymbol{\alpha}). \tag{3.17}$$

■

This theorem gives us an easily verifiable criterion that allows us to check whether a function is generalized convex. All that is necessary to do is to solve two consecutive constraint maximization problems. There exists another way to characterize the set of $u-$convex functions.

Proposition 63 *Function $s(\cdot)$ is $u-$convex if and only if there exists a non-empty set $Y \subset R^n \times R^*$ such that*

$$s(\boldsymbol{\alpha}) = \sup_{(\mathbf{x},\mathbf{t})\in Y}(u(\boldsymbol{\alpha}, \mathbf{x}) - t). \tag{3.18}$$

Proof. First, suppose (3.18) holds. Recall that

$$s^*(\mathbf{x}) = \sup_{\boldsymbol{\alpha}\in\Omega}(u(\boldsymbol{\alpha}, \mathbf{x}) - s(\boldsymbol{\alpha})). \tag{3.19}$$

According to (3.18) for all $(\mathbf{x}, t) \in Y$

$$t \geq u(\boldsymbol{\alpha}, \mathbf{x}) - s(\boldsymbol{\alpha}), \tag{3.20}$$

therefore

$$t \geq s^*(\mathbf{x}) \tag{3.21}$$

and

$$u(\boldsymbol{\alpha}, \mathbf{x}) - t \leq u(\boldsymbol{\alpha}, \mathbf{x}) - s^*(\mathbf{x}) \tag{3.22}$$

for all $(\mathbf{x}, t) \in Y$. Let X be projection of Y on R^n. Then

$$s(\boldsymbol{\alpha}) = \sup_{(\mathbf{x},\mathbf{t})\in Y}(u(\boldsymbol{\alpha}, \mathbf{x}) - t) \leq \sup_{\mathbf{x}\in X}(u(\boldsymbol{\alpha}, \mathbf{x}) - s^*(\mathbf{x})). \tag{3.23}$$

But

$$\sup_{\mathbf{x}\in X}(u(\boldsymbol{\alpha}, \mathbf{x}) - s^*(\mathbf{x})) \leq \sup_{\mathbf{x}\in R^n}(u(\boldsymbol{\alpha}, \mathbf{x}) - s^*(\mathbf{x})) = s^{**}(\boldsymbol{\alpha}), \tag{3.24}$$

therefore $s(\boldsymbol{\alpha}) \leq s^{**}(\boldsymbol{\alpha})$. On the other hand, $s^{**}(\boldsymbol{\alpha}) \leq s(\boldsymbol{\alpha})$ for any $s(\cdot)$, therefore $s(\boldsymbol{\alpha}) = s^{**}(\boldsymbol{\alpha})$ and $s(\cdot)$ is $u-$convex.

Now, suppose that

$$s(\boldsymbol{\alpha}) = \sup_{\mathbf{x}\in R^n} (u(\boldsymbol{\alpha}, \mathbf{x}) - t(\mathbf{x})), \tag{3.25}$$

for some function $t(\cdot) : R^n \to R^*$. Let T be the image of R^n under $t(\cdot)$ ($T = t(R^n)$) and define $Y = R^n \times T$. Then$s(\boldsymbol{\alpha}) = \sup_{(\mathbf{x},\mathbf{t})\in Y} (u(\boldsymbol{\alpha}, \mathbf{x}) - t)$. ■

The final result we are going to prove is this section states that the upper envelope of the set of $u-$convex functions is $u-$convex. It generalizes the well-known result for the convex functions.

Proposition 64 *Let functions $s(\cdot, \beta)$ be $u-$convex for every $\beta \in \Theta$. Then*

$$s(\boldsymbol{\alpha}) = \sup_{\beta\in\Theta} s(\alpha, \beta) \tag{3.26}$$

is $u-$convex.

Proof. Since each $s(\cdot, \beta)$ is $u-$convex for there exits $t(\mathbf{x}, \beta)$ such that

$$s(\cdot, \beta) = \sup_{\mathbf{x}\in R^n} (u(\boldsymbol{\alpha}, \mathbf{x}) - t(\mathbf{x},\beta)). \tag{3.27}$$

But then

$$s(\boldsymbol{\alpha}) = \sup_{\beta\in\Theta} s(\alpha, \beta) = \sup_{\beta\in\Theta} \sup_{\mathbf{x}\in R^n} (u(\boldsymbol{\alpha}, \mathbf{x}) - t(\mathbf{x},\beta)). \tag{3.28}$$

Since $u(\cdot, \cdot)$ does not depend on β

$$s(\boldsymbol{\alpha}) = \sup_{\mathbf{x}\in R^n} (u(\boldsymbol{\alpha}, \mathbf{x}) - \inf_{\beta\in\boldsymbol{\Theta}} t(\mathbf{x},\beta)) = \sup_{\mathbf{x}\in R^n} (u(\boldsymbol{\alpha}, \mathbf{x}) - t(\mathbf{x})), \tag{3.29}$$

where

$$t(\mathbf{x}) = \inf_{\beta\in\boldsymbol{\Theta}} t(\mathbf{x},\beta). \tag{3.30}$$

Therefore, $s(\cdot)$ is $u-$convex. ■

3.2 Generalized Convexity and Cyclic Monotonicity

In this Section I am going to define notions of a subdifferential of a function and $u-$cyclically monotonic vector field. I will prove prove that subdifferentials of the $u-$convex functions are $u-$cyclically monotonic and for any $u-$cyclically monotonic vector field $\mathbf{b}$ there exists a $u-$convex function such that $\mathbf{b}$ belongs to its subdifferential.

Let $s(\cdot) : \Omega \to R$ be a $u-$convex function, where $u(\cdot, \cdot) : \Omega \times R^n \to R$ is a differentiable function. Define a subdifferential of $s(\cdot)$ at point α_0 in the following way:

Definition 65 *The subdifferential of a $u-$convex function $s(\cdot)$ at point α_0 is the following set:*

$$\partial s(\boldsymbol{\alpha}_0) = \{\mathbf{x} \in R^n : s(\boldsymbol{\alpha}) - s(\boldsymbol{\alpha}_0) \geq u(\boldsymbol{\alpha}, \mathbf{x}) - u(\boldsymbol{\alpha}_0, \mathbf{x})\} \text{ for } \forall \boldsymbol{\alpha} \in \Omega. \tag{3.31}$$

Assume that if $s(\cdot)$ is differentiable at $\boldsymbol{\alpha}_0$ and let

$$X = \{\mathbf{x} \in R^n : \nabla s(\boldsymbol{\alpha}_0) = \nabla_{\boldsymbol{\alpha}} u(\boldsymbol{\alpha}, \mathbf{x})\}. \tag{3.32}$$

Then $\partial s(\boldsymbol{\alpha}_0) \subset X$. Now let $\mathbf{x} : \Omega \to R^n$ be a vector field. Next, we are going to introduce the notion of cyclically monotonicity of a vector field.

Definition 66 *Vector field $\mathbf{x}(\cdot)$ is called $u-$cyclically monotonic if for any n points $\boldsymbol{\alpha}_1, ..., \boldsymbol{\alpha}_n \in \Omega$*

$$\sum_{i=1}^{n} (u(\boldsymbol{\alpha}_{i+1}, \mathbf{x}_i) - u(\boldsymbol{\alpha}_i, \mathbf{x}_i)) \leq 0, \tag{3.33}$$

where $\boldsymbol{\alpha}_{n+1} = \boldsymbol{\alpha}_1$, and $\mathbf{x}_i = \mathbf{x}(\boldsymbol{\alpha}_i)$.

Let us start with proving the following result:

Lemma 67 *Let $\Omega \subset R^m$ be an open, convex set. Suppose that $u(\cdot,\cdot)$ is continuously differentiable on $\Omega \times R^n$ and vector field $\mathbf{x}(\cdot)$ is continuously differentiable and $u-$cyclically monotonic on Ω. Then the vector field $\mathbf{b}(\cdot)$ defined by*

$$b_i(\boldsymbol{\alpha}) = \frac{\partial u}{\partial \alpha_i}(\boldsymbol{\alpha}, \mathbf{x}(\boldsymbol{\alpha})) \tag{3.34}$$

is conservative.

Proof. According to Corollary 19 to prove the Lemma we should establish that for any smooth closed curve Γ

$$\oint_{\Gamma} \mathbf{b}(\boldsymbol{\alpha}) \mathbf{d}\boldsymbol{\alpha} = 0. \tag{3.35}$$

It is sufficient to prove that for any smooth closed curve Γ

$$\oint_{\Gamma} \mathbf{b}(\boldsymbol{\alpha}) \mathbf{d}\boldsymbol{\alpha} \leq 0, \tag{3.36}$$

since if Γ is a smooth closed path so is Γ^- (path Γ travelled in the opposite direction) and

$$\oint_{\Gamma^-} \mathbf{b}(\boldsymbol{\alpha}) \mathbf{d}\boldsymbol{\alpha} = -\oint_{\Gamma} \mathbf{b}(\boldsymbol{\alpha}) \mathbf{d}\boldsymbol{\alpha}. \tag{3.37}$$

Take a smooth closed curve Γ and let $\boldsymbol{\alpha}_1, ..., \boldsymbol{\alpha}_n, \boldsymbol{\alpha}_{n+1} = \boldsymbol{\alpha}_1 \in \Gamma$. Then cyclic monotonicity implies that

$$\sum_{i=1}^{n} (u(\boldsymbol{\alpha}_{i+1}, \mathbf{x}_i) - u(\boldsymbol{\alpha}_i, \mathbf{x}_i)) = \sum_{i=1}^{n} \int_0^1 \frac{\partial u}{\partial \alpha_i}(\boldsymbol{\alpha}_i + t(\boldsymbol{\alpha}_{i+1} - \boldsymbol{\alpha}_i), \mathbf{x}_i)(\boldsymbol{\alpha}_{i+1} - \boldsymbol{\alpha}_i) dt \leq 0. \tag{3.38}$$

But

$$\lim_{\max_i \|\boldsymbol{\alpha}_{i+1} - \boldsymbol{\alpha}_i\| \to 0} \sum_{i=1}^{n} \int_0^1 \frac{\partial u}{\partial \alpha_i}(\boldsymbol{\alpha}_i + t(\boldsymbol{\alpha}_{i+1} - \boldsymbol{\alpha}_i), \mathbf{x}_i)(\boldsymbol{\alpha}_{i+1} - \boldsymbol{\alpha}_i) dt = \oint_{\Gamma} \mathbf{b}(\boldsymbol{\alpha}) \mathbf{d}\boldsymbol{\alpha}, \tag{3.39}$$

which establishes the result. ■

The main result of this Section establishes the connection between $u-$cyclically monotonic vector fields and subdifferentials of $u-$convex functions. Let us start by proving the following lemma.

Lemma 68 *Vector field $\mathbf{x}(\boldsymbol{\alpha})$ is $u-$cyclically monotonic if there exists $u-$convex function $s(\cdot)$ such that $\mathbf{x}(\boldsymbol{\alpha}) \in \partial s(\boldsymbol{\alpha})$ for $\forall \boldsymbol{\alpha} \in \Omega$.*

Proof. Let us assume that $\mathbf{x}(\boldsymbol{\alpha}) \in \partial s(\boldsymbol{\alpha})$ for some $u-$convex function $s(\cdot)$ then vector field $\mathbf{x}$ is $u-$cyclically monotonic, since in that case

$$\sum_{i=1}^{n} (u(\boldsymbol{\alpha}_{i+1}, \mathbf{x}_i) - u(\boldsymbol{\alpha}_i, \mathbf{x}_i)) \leq \sum_{i=1}^{n} (s(\boldsymbol{\alpha}_{i+1}) - s(\boldsymbol{\alpha}_i)) = 0. \tag{3.40}$$

Therefore, $\mathbf{x}(\boldsymbol{\alpha})$ is $u-$cyclically monotonic.

Now assume that $\mathbf{x}(\boldsymbol{\alpha})$ is $u-$cyclically monotonic. Fix $\boldsymbol{\alpha}_0 \in \Omega$ and let

$$C_n = \{\boldsymbol{\alpha}_0, \boldsymbol{\alpha}_1, ..., \boldsymbol{\alpha}_{n+1} = \boldsymbol{\alpha}\} \tag{3.41}$$

be an $(n+1)-$element chain connecting points $\boldsymbol{\alpha}_0$ and $\boldsymbol{\alpha}$. Define $s(\cdot)$ by

$$s(\boldsymbol{\alpha}) = s(\boldsymbol{\alpha}_0) + \sup_n \sup_{C_n} (\sum_{i=0}^{n} (u(\boldsymbol{\alpha}_{i+1}, \mathbf{x}_i) - u(\boldsymbol{\alpha}_i, \mathbf{x}_i))), \tag{3.42}$$

where $s(\boldsymbol{\alpha}_0)$ is arbitrary. Note that each term under the sign of double supremum has a form

$$u(\boldsymbol{\alpha}, \overline{\mathbf{x}}) + \overline{t} \tag{3.43}$$

for some fixed $\overline{\mathbf{x}}$ and $\overline{t}$. Therefore, each term is $u-$convex and hence, so is $s(\boldsymbol{\alpha})$ as an upper envelope of $u-$convex functions. Moreover, taking just a one element chain connecting points $\boldsymbol{\alpha}_0$ and $\boldsymbol{\alpha}$ we arrive at

$$s(\boldsymbol{\alpha}) - s(\boldsymbol{\alpha}_0) \geq u(\boldsymbol{\alpha}, \mathbf{x}_0) - u(\boldsymbol{\alpha}_0, \mathbf{x}_0). \tag{3.44}$$

Therefore, $\mathbf{x}_0 \in \partial s(\boldsymbol{\alpha}_0)$. ■

3.3 Examples

Example 69 *Assume $\Omega = (0,1)$ and $u : \Omega \times R^2 \to R$ be defined by*

$$u(\alpha, x) = \alpha x_1 + \alpha^{3/2} x_2.$$

Consider a function $s(\alpha) = \alpha$. Then

$$s^*(\mathbf{x}) = \max_{\alpha \in [0,1]} (\alpha x_1 + \alpha^{3/2} x_2 - \alpha).$$

Since the objective is convex, maximum is achieved at the one of the end points. Therefore,

$$s^*(\mathbf{x}) = \max(x_1 + x_2 - 1, 0).$$

Now

$$s^{**}(\alpha) = \max_{x \in R^2} (\alpha x_1 + \alpha^{3/2} x_2 - s^*(x)).$$

The maximum is obtained at point

$$x_1 = 1, \ x_2 = 0$$

and the value of the maximum is α. Hence $s(\cdot)$ is $u-$convex.

Example 70 *Assume $\Omega = (0,1)$ and $u : \Omega \times R^2 \to R$ be defined by*

$$u(\alpha, x) = \alpha^{5/4} x_1 + \alpha^{3/2} x_2.$$

Consider a function $s(\alpha) = \alpha$. Then going through the calculations similar to the previous example

$$s^*(x) = \max(x_1 + x_2 - 1, 0)$$

and

$$s^{**}(\alpha) = \max_{x \in R^2} (\alpha^{5/4} x_1 + \alpha^{3/2} x_2 - s^*(x)) = \alpha^{5/4} \neq \alpha.$$

Therefore, $s(\cdot)$ is not $u-$convex.

Example 71 *In this example I am going to describe two classes of u−convex functions. First, assume*

$$u(\boldsymbol{\alpha},\mathbf{x}) = \min_{i=1,\dots,n} \alpha_i x_i.$$

The class of u−convex functions in this case is the class of lower semicontinuous convex along rays functions (Abasov and Rubinov, 1994). This class is rather broad. It includes convex functions and homogenous of degree β functions, provided $\beta \geq 1$. It is straightforward to see that u−convex functions functions are convex along the rays, since across any ray

$$\alpha_i = t\xi_i \tag{3.45}$$

for some vector $\boldsymbol{\xi}$ of a unit length

$$u(\boldsymbol{\alpha},\mathbf{x}) = t \min_{i=1,\dots,n} \alpha_i \xi_i \tag{3.46}$$

is linear in t. Therefore, convexity along rays is necessary. For a proof of sufficiency the reader is referred to Abasov and Rubinov (1994).

Example 72 *Let $n = m + 1$ and*

$$u(\boldsymbol{\alpha},\mathbf{x}) = \sum_{i=1}^{m} \alpha_i x_i - \frac{1}{2}\left\|\alpha\right\|^2 x_{m+1}.$$

Then any continuous function on Ω is u−convex. For a proof, see Kutateladze and Rubinov (1972)

The last example shows that the class of u−convex functions can be extremely broad. We will return to this point in the second part of the book.

3.4 Exercises

1. Prove that for any function $s(\cdot)$

$$s^{**} \leq s. \tag{3.47}$$

2. Let $\Omega = (0,1) \times (0,1)$ and

$$u(\boldsymbol{\alpha},\mathbf{x}) = \alpha_1 x_1 + \alpha_1\alpha_2 x_2 + \alpha_2 x_3. \tag{3.48}$$

Find u−Fenchel conjugates and u−Fenchel biconjugates of the following functions:

a). $s(\boldsymbol{\alpha}) = \alpha_1^2 + \alpha_2^2$
b). $s(\boldsymbol{\alpha}) = \alpha_1 + 2\alpha_2$
c). $s(\boldsymbol{\alpha}) = \alpha_1^{3/2}\alpha_2^{1/2} + \alpha_1^2$
d). $s(\boldsymbol{\alpha}) = \alpha_1^{1/2} + \alpha_2.$

3.5 Bibliographic Notes

Abstract convex analysis is a relatively new field of the mathematical research. A lot of interesting results can be still found only in the journal articles. However, there are already some good books on the subject. The best expositions of the area are probably Singer (1997) and Rubinov (2000). The latter contains numerous examples of classes of $u-$convex functions. These examples are of the great interest to a student of the screening models and of a more broad field of the multidimensional mechanism design. Some of them were given in this chapter. I also make a use of them in the second part of the book. An excellent exposition of subdifferentiability of convex functions and cyclic monotonicity of the subdifferential can be found in Rockafeller (1997).

4

Calculus of Variations and the Optimal Control

In this chapter we are going to study the problem

$$\max_{\mathbf{u}\in V} L(\mathbf{u}),$$

where V is some normed vector space (usually a Banach space) and $L(\cdot)$ is functional, i. e. a mapping from V into R.

4.1 Banach Spaces and Polish Spaces

Let V be a vector space over R.

Definition 73 *A function $\|\cdot\| : V \to R$ is called a norm if it satisfies the following properties:*

1. $\|\mathbf{x}\| \geq 0$ *for* $\forall \mathbf{x} \in V$ *. Moreover,* $\|\mathbf{x}\| = 0$ *if and only if* $\mathbf{x} = \mathbf{0}$*;*
2. $\|\boldsymbol{\lambda}\mathbf{x}\| = |\lambda| \, \|\mathbf{x}\|$ *, for* $\forall \mathbf{x} \in V$ *and* $\forall \lambda \in R$*;*
3. $\|\mathbf{x}+\mathbf{y}\| \leq \|\mathbf{x}\| + \|\mathbf{y}\|$ *, for* $\forall \mathbf{x}, \mathbf{y} \in V$ *.*

Vector space V together with norm $\|\cdot\|$ is called a normed vector space. It is usually referred to as $(V, \|\cdot\|)$ or simply V if no confusion is possible. Let us consider some examples of normed spaces.

Example 74 *Let $V = R^n$ and for $\forall \mathbf{x} \in V$ define the norm by*

$$\|\mathbf{x}\| = \sqrt{\sum_{i=1}^{n} x_i^2}. \tag{4.1}$$

It is easy to check that (4.1) is indeed a norm. This norm is known is the Euclidean norm.

Example 75 *Let $\Omega \subset R^n$ be a compact set and V the set of continuous functions from Ω into the real line. For $\forall h \in V$, define the norm by*

$$\|h\| = \max_{\mathbf{x}\in\Omega} |h(\mathbf{x})| . \tag{4.2}$$

Note that since $h(\cdot)$ is continuous and Ω is compact the maximum in (4.2) is well-defined. It is easy to check that (4.2) is indeed a norm. This norm is known as the norm of uniform convergence and space of continuous real valued functions endowed with this norm is usually denoted as $C(\Omega)$.

Example 76 *Let $\Omega \subset R^n$ be a Borel set and let $\Lambda(\Omega)$ denote the set of Lebesgue measurable function from Ω into the real line. Call two functions $h_1, h_2 \in \Lambda$ equivalent and write*

$$h_1 \sim h_2 \tag{4.3}$$

if the set

$$D = \{\mathbf{x} \in\Omega : h_1(x) \neq h_2(x)\} \tag{4.4}$$

has Lebesgue measure zero. Relation $\sim$ partitions set Ω into the classes of equivalent elements. Define the factor space $\Lambda/\sim$ as the space whose elements are the equivalence classes. Fix some $p \geq 1$ and take some $f \in \Lambda/\sim$. Assume that $\exists f_0 \in f$ such that

$$\int_\Omega |f_0(\mathbf{x})|^p \, \mathbf{dx} < \infty. \tag{4.5}$$

It is easy to show that then it holds for $\forall f_0 \in f$ and the value of the integral does not depend of the choice of f_0. Now let

$$L^p(\Omega) = \{f \in \Lambda/\sim : \exists f_0 \in f : \int_\Omega |f_0(\mathbf{x})|^p \, \mathbf{dx} < \infty\} \tag{4.6}$$

and for $\forall f \in L^p(\Omega)$ define the norm by

$$\|f\| = (\int_\Omega |f_0(\mathbf{x})|^p \, \mathbf{dx})^{1/p}. \tag{4.7}$$

It is easy to check that (4.7) is indeed a norm.

The last example is of a particular importance for us in the future. We will see that it allows us to introduce Sobolev spaces, which are the natural spaces in which the solutions to the screening models live.

Our next objective is to define the notion of completeness of a normed space. We will start with definitions of a converging and a fundamental sequences in a normed space.

Definition 77 *We say that sequence* $\{x_n\}_{n=0}^{\infty}$ *of the elements of normed space* V *converges to point* $x \in V$ *and write*

$$\lim_{n \to \infty} x_n = x \tag{4.8}$$

if

$$\forall \varepsilon > 0 \ \exists n_0 \in N \text{ such that } \forall n > n_0, \ \|x_n - x\| < \varepsilon. \tag{4.9}$$

Definition 78 *A sequence* $\{x_n\}_{n=0}^{\infty}$ *of the elements of normed space* V *is called fundamental if*

$$\forall \varepsilon > 0 \ \exists n_0 \in N \text{ such that } \forall n, m > n_0, \ \|x_m - x_n\| < \varepsilon. \tag{4.10}$$

It is easy to check that any converging sequence is fundamental. The reverse is, however, not true in general. Consider the following example. Let $V = Q$ is the set of rational numbers with norm defined by $\|x\| = |x|$. Let $\{x_n\}_{n=0}^{\infty}$ be the sequence of decimal approximation to $\sqrt{2}$ up to n digits after the decimal point, i. e. $x_0 = 1$, $x_1 = 1.4$, $x_2 = 1.41$, etc. Then for any $m, n > n_0$:

$$\|x_m - x_n\| < 10^{-n_0+1}, \tag{4.11}$$

therefore the sequence is fundamental, but it does not converge to any rational number (since it converges to $\sqrt{2}$, which is irrational). If we add to Q limits of all fundamental sequences we will get the set R of real numbers, which has a property that every fundamental sequence is converging.

Definition 79 *A normed space is called complete if every fundamental sequence is converging. A complete, linear, normed space is called a Banach space.*

Suppose we have a normed space. Then we can introduce a distance function $d : V \times V \to R$ as

$$d(x, y) = \|x - y\|. \tag{4.12}$$

It is easy to check that this function has the following properties:

$$\begin{aligned} &1. \ d(x,y) \ \geq \ 0 \text{ for } \forall x, y \in V. \text{ Moreover, } d(x,y) = 0 \Leftrightarrow x = y; \\ &2. \ d(x,y) \ = \ d(y,x), \text{for } \forall x, y \in V; \\ &3. \ d(x,z) \ \leq \ d(x,y) + d(y,z), \text{ for } \forall x, y, z \in V. \end{aligned} \tag{4.13}$$

Instead of introducing $d(\cdot, \cdot)$ by (4.12) we could have introduced it axiomatically as any function satisfying (4.13). That leads us to the following definition:

Definition 80 *A set* X *together with a function* $d : X \times X \to R$ *satisfying (4.13) is called a metric space.*

We saw already that any normed vector space can be considered as a metric space. Metrics, however, need not come from a norm. Moreover, metric does not require any linear structure and can be defined on any set, not only on a vector space. Next, we are going to develop notions of convergence and completeness for metric spaces. They parallel similar definitions for the normed spaces.

Definition 81 *We say that sequence* $\{x_n\}_{n=0}^{\infty}$ *of the elements of metric space* X *converges to point* $x \in X$ *and write*

$$\lim_{n \to \infty} x_n = \mathbf{x} \tag{4.14}$$

if

$$\forall \varepsilon > 0 \ \exists n_0 \in N \text{ such that } \forall n > n_0, \ d(x_n, x) < \varepsilon. \tag{4.15}$$

Definition 82 *A sequence* $\{x_n\}_{n=0}^{\infty}$ *of the elements of metric space* X *is called fundamental if*

$$\forall \varepsilon > 0 \ \exists n_0 \in N \text{ such that } \forall n, m > n_0, \ d(x_n, x_m) < \varepsilon. \tag{4.16}$$

Most metric spaces one encounters in economics has an additional property called separability. To define the notion of a separable space, let us start with following definitions.

Definition 83 *A set* X *is called countable is there is a one to one correspondence between* X *and the set of natural numbers* N.

Definition 84 *The closure of set* X, *denoted by* $cl(X)$, *is the minimal in the sense of the set-theoretic inclusion closed set that includes* X.

Definition 85 *A metric space* X *is called separable if there exists a countable subset* $Y \subset X$ *such that* $X = cl(Y)$.

Finally, I will give the following definition:

Definition 86 *A metric space is called complete if every fundamental sequence is converging. A complete separable metric space is called a Polish space.*

Using definition of metric one can straightforwardly generalize notions of an open set, a closed set, and a continuous function for metric spaces. They will be verbatim the same as in the usual calculus. Let us, for example, define an open set. First, define an open ball with radius $\varepsilon > 0$ and center x as

$$B_\varepsilon(x) = \{y \in X : d(x, y) < \varepsilon\} \tag{4.17}$$

and the closed ball as

$$\overline{B_\varepsilon(x)} = \{y \in X : d(x, y) \leq \varepsilon\}. \tag{4.18}$$

Definition 87 *A set $U \subset X$ is called open if for any $x \in U$ there exists $\varepsilon > 0$ such that $B_\varepsilon(x) \subset U$. A set $F \subset X$ is called closed if its complement is open.*

It is left as an exercise to the reader to proof that the open ball is an open set and the closed ball a closed set and to define a notion of a continuous function from one metric space into another.

4.2 Hilbert Spaces

Let X be a linear space over R , and $\langle \cdot, \cdot \rangle : X \times X \to R$ be a binary operation on it satisfying the following properties:
(i). $\langle a, b\rangle = \langle b, a\rangle$ for $\forall a, b \in X$
(ii). $\langle \lambda a + \mu c, b\rangle = \lambda \langle a, b\rangle + \mu \ \langle c, b\rangle$ for $\forall a, b, c \in X$, $\forall \lambda, \mu \in R$
(iii). $\langle a, a\rangle \geq 0$ for $\forall a \in X$ and $\langle a, a\rangle = 0$ if and only if $a = 0$.

Definition 88 *A linear space X over R together with a binary operation $\langle \cdot, \cdot \rangle$ satisfying (i)-(iii) is called a Euclidean space. Operation $\langle \cdot, \cdot \rangle$ is called an inner product.*

Definition 89 *Let $(X, \langle \cdot, \cdot \rangle)$ be a Euclidean space over R and $Y \subset X$ is such that Y is also a linear space. Then $(Y, \langle \cdot, \cdot \rangle)$ is called a Euclidean subspace of $(X, \langle \cdot, \cdot \rangle)$.*

A fundamental property of the inner product is given by Cauchy-Schwarz inequality.

Lemma 90 *Let X be a Euclidean space. Then for any $a, b \in X$*

$$|\langle a, b\rangle| \leq \sqrt{\langle a, a\rangle} \cdot \sqrt{\langle b, b\rangle}. \tag{4.19}$$

Proof. By property (iii)

$$\langle a - \lambda b, a - \lambda b\rangle \geq 0 \tag{4.20}$$

for any $\lambda \in R$. Using (i)-(ii) one can transform (4.20) to read

$$\langle b, b\rangle \lambda^2 - 2\langle a, b\rangle \lambda + \langle a, a\rangle \geq 0 \tag{4.21}$$

for any $\lambda \in R$. But a quadratic polynomial in λ with a nonnegative coefficient before λ^2 is always nonnegative if and only if its discriminant, D, is non-positive, where

$$D = 4\langle a, b\rangle^2 - 4\langle a, a\rangle \langle b, b\rangle. \tag{4.22}$$

Therefore, (4.19) follows. ∎

Using Cauchy-Schwarz inequality one can easily prove that function $\|\cdot\|$ defined by

$$\|a\| = \sqrt{\langle a, a \rangle} \tag{4.23}$$

is a norm. I leave it as an exercise for the reader. Let me give some examples of the Euclidean spaces.

Example 91 $R^n = \{(x_1, ..., x_n) : x_i \in R\}$. *For any* $\mathbf{a}, \mathbf{b} \in R^n$ *define*

$$\langle \mathbf{a}, \mathbf{b} \rangle \equiv \mathbf{a} \cdot \mathbf{b} = \sum_{i=1}^{n} a_i b_i. \tag{4.24}$$

It is easy to check that (i)-(iii) are satisfied, therefore R^n *endowed with the inner product (4.24) is a Euclidean space.*

Example 92 *Space* $L^2(\Omega)$, *where* $\Omega \in R^n$ *is a Borel set and the inner product*

$$\langle f, g \rangle = \int_{\Omega} f(\mathbf{x}) g(\mathbf{x}) \mathbf{dx}. \tag{4.25}$$

Note that by Cauchy-Schwarz inequality

$$\left| \int_{\Omega} f(\mathbf{x}) g(\mathbf{x}) \mathbf{dx} \right| \leq \sqrt{\int_{\Omega} f^2(\mathbf{x}) \mathbf{dx}} \sqrt{\int_{\Omega} g^2(\mathbf{x}) \mathbf{dx}}, \tag{4.26}$$

therefore integral (4.25) is converging for any $f, g \in L^2(\Omega)$.

A particular important examples of Euclidean spaces are the so-called Hilbert spaces.

Definition 93 *Let* $(X, \langle \cdot, \cdot \rangle)$ *be a Euclidean space and assume that the normed space* $(X, \|\cdot\|)$, *where norm* $\|\cdot\|$ *is defined by (4.23) is complete. Then* $(X, \langle \cdot, \cdot \rangle)$ *is called a Hilbert space.*

Exercise 3 after this chapter asks the reader to prove that R^n and $L^p(\Omega)$ for closed Ω and $p > 1$ are Banach spaces. This will imply that R^n and $L^2(\Omega)$ for closed Ω are Hilbert spaces.

4.3 Dual Space for a Normed Space and a Hilbert Space

In this Section I am going to define a dual space of a normed linear space. Let us start with defining a linear mapping between two linear spaces.

Definition 94 *Let V and U be linear spaces. A mapping $f : V \to U$ is called linear if*

$$f(\lambda x + \mu y) = \lambda f(x) + \mu f(y) \tag{4.27}$$

for $\forall x, y \in V$ and $\forall \lambda, \mu \in R$. If $U = R$ or C the mapping is usually called a functional, otherwise it is called an operator.

Definition 95 *A functional $f : V \to R$ is called continuous if for any sequence $\{x_n\}_{n=0}^{\infty}$ of the elements of normed space V*

$$\lim_{n\to\infty} x_n = x \tag{4.28}$$

implies

$$\lim_{n\to\infty} f(x_n) = f(x). \tag{4.29}$$

Define dual of the normed linear space V as the set of all linear continuous functionals on V. I will denote this space V^* and endow it with a norm

$$\|f\| = \sup_{x, \|x\|=1} |f(x)|. \tag{4.30}$$

Theorem 96 *Equation (4.30) defines a norm on V^*.*

Proof. First, note that defining

$$\begin{aligned} (f+g)(x) &= f(x) + g(x) && (4.31) \\ (\lambda f)(x) &= \lambda f(x) && (4.32) \end{aligned}$$

for all $x \in V$ and defining the zero functional $\mathbf{0}$ as one that identically vanishes on V makes V^* a linear space. Let $\|f\|$ be defined by (4.30). Then, obviously $\|f\| \geq 0$. Moreover, $\|f\| = 0$ implies that $f(x) = 0$ for $\forall x \in V$ such that $\|x\| = 1$. Take any $x \in V$. If $x = 0$ then take any $y \in V$ and write

$$f(0) = f(0y) = 0f(y) = 0, \tag{4.33}$$

where we used linearity of f. If $x \neq 0$ let $y = x/\|x\|$ and (again, using linearity of f)

$$f(x) = \|x\| f(y) = 0, \tag{4.34}$$

the last equality follows from the fact $\|y\| = 1$. Since $\|\mathbf{0}\| = 0$ we proved that $\|f\| = 0$ if and only if $f = \mathbf{0}$. Note that for $\forall \lambda$

$$\|\lambda f\| = \sup_{x, \|x\|=1} |\lambda f(x)| = |\lambda| \sup_{x, \|x\|=1} |f(x)| = |\lambda| \|f\|. \tag{4.35}$$

Finally, for any $f, g \in V^*$

$$\|f + g\| = \sup_{x, \|x\|=1} |f(x) + g(x)|. \tag{4.36}$$

Since for real numbers

$$|f(x)+g(x)| \leq |f(x)|+|g(x)|, \tag{4.37}$$

and

$$\sup_{x,\|x\|=1}(|f(x)|+|g(x)|) \leq \sup_{x,\|x\|=1}(|f(x)|)+\sup_{x,\|x\|=1}(|g(x)|), \tag{4.38}$$

one obtains

$$\|f+g\| \leq \|f\|+\|g\|. \tag{4.39}$$

Therefore, (4.30) defines a norm. ∎

It turns out that the dual of a linear space with norm (4.30) is a Banach space. I will not prove it here and refer an interested reader to Abraham, Mardsen, and Raitu (1988). The set of linear functionals on X has a particular simple characterization if X is a Hilbert space.

Theorem 97 *Let $(X,\langle\cdot,\cdot\rangle)$ be a Hilbert space and $f: X \rightarrow R$ be a linear functional. Then there exists a unique $h_f \in X$ such that $\forall x \in X$*

$$f(x)=\langle h_f, x\rangle. \tag{4.40}$$

This theorem states that the dual of a Hilbert space can be naturally identified with the space itself. I will not give the proof in the general case, since it requires some recourse in the axiomatic set theory (for example, it uses the Axiom of Choice). I will, however, provide the proof in the finite dimensional case, which contains all the main ideas of the proof in the general case.

Definition 98 *Hilbert space $(X,\langle\cdot,\cdot\rangle)$ is called finite-dimensional if there exists a finite set $\{e_1,...,e_n\}$ of its elements such that any $x \in X$ is uniquely representable as*

$$x=\sum_{i=1}^{n}\alpha_i e_i. \tag{4.41}$$

Set of elements $\{e_1,...,e_n\}$ is called a basis of X.

Definition 99 *Basis $\{e_1,...,e_n\}$ is called orthonormal if*

$$\langle e_i, e_j\rangle=\delta_{ij} \equiv \begin{cases} 1, & \text{if } i=j \\ 0, & \text{if } i \neq j \end{cases}. \tag{4.42}$$

Every finite dimensional Hilbert space has an orthonormal basis. Indeed, start with any basis $\{e_1,...,e_n\}$ and define inductively

$$f_1 = \frac{e_1}{\|e_1\|}, \tag{4.43}$$

$$f_2' = e_2-\langle f_1, e_2\rangle f_1,\ f_2=\frac{f_2'}{\|f_2\|}, \tag{4.44}$$

$$f_k' = e_k-\sum_{i=1}^{k-1}\langle f_i, e_k\rangle f_i,\ f_k=\frac{f_k'}{\|f_k\|}. \tag{4.45}$$

The process is known as Schmidt orthogonalization procedure. It is easy to check that $\{f_1, ..., f_n\}$ is an orthonormal basis. Now we are ready to prove the Theorem for a finite dimensional Hilbert space.
Proof. Given a linear functional $f : X \to R$ define $h \in X$ by

$$h_f = \sum_{i=1}^{n} f(e_i)e_i, \tag{4.46}$$

then $\forall x \in X$. Then

$$\langle h_f, x\rangle = \langle \sum_{i=1}^{n} f(e_i)e_i, \sum_{j=1}^{n} \alpha_j e_j\rangle = \sum_{i=1}^{n} \alpha_i f(e_i) = f(x). \tag{4.47}$$

Proof of uniqueness is trivial and is left as an exercise to the reader. ■

The only complication that arises in the infinite dimensional case is the necessity to check that X possesses a basis, i. e. an infinite set of elements $\{e_\alpha\}_{\alpha \in I}$ for some infinite I such that an appropriately defined analog of (4.41) holds. This is where the choice axiom is necessary. The rest of the proof remains the same.

4.4 Frechet Derivative of a Mapping between Normed Spaces

For a differentiable function $f : R^n \to R$ its gradient at point $\mathbf{x}_0 \in R^n$ defines a linear functional $h(\cdot)$ on R^n by

$$h(z) = \nabla f(\mathbf{x}_0) \cdot \mathbf{z}. \tag{4.48}$$

For a differentiable mapping $\phi : R^n \to R^m$ its Jacobi matrix evaluated at $\mathbf{x}_0 \in R^n$, $J(\mathbf{x}_0)$ defines a linear operator A from R^n into R^m defined by

$$A\mathbf{z} = J(\mathbf{x}_0)\mathbf{z}. \tag{4.49}$$

We will use this observations to define a derivative for a mapping between arbitrary linear spaces.

Let V and U be two linear spaces and $L(U, V)$ be the set of all linear operators acting from V into U.

Definition 100 *We say that mapping $\varphi : U \to V$ is o−small of x and write*

$$\varphi(x) = o(x) \tag{4.50}$$

if

$$\lim_{x \to 0} \frac{||\varphi(x)||}{||x||} = 0. \tag{4.51}$$

Definition 101 *Mapping* $\varphi : U \to V$ *is said to be differentiable at* $x_0 \in U$ *if* $\exists A \in L(U, V)$ *such that*

$$\varphi(x_0 + h) = \varphi(x_0) + Ah + o(h). \tag{4.52}$$

Mapping A *is called Frechet derivative of mapping* φ *at* x_0 *and denoted* $D\phi(x_0)$*. If it is a continuous mapping,* φ *is called continuously differentiable.*

It is clear that if a mapping is differentiable at a point it is continuous at it. Many other results from the finite-dimensional calculus (e. g. the chain rule) generalize to the Frechet calculus. The proof is almost verbatim the same as in the finite-dimensional calculus and I will not give it here. An interested reader is referred to Abraham, Mardsen, Ratiu (1988). Instead, let us restrict our attention to the special case when $V = R$ (the mapping is a functional), which is of a particular importance for our applications.

Definition 102 *Point* $x_0 \in U$ *is called a local maximum (minimum) of the functional* $\phi : U \to R$ *if* $\exists \varepsilon > 0$ *such that* $\phi(x) \leq \phi(x_0)$ $(\phi(x) \geq \phi(x_0))$ *for any* $x \in U$ *such that* $\|x - x_0\| < \varepsilon$*.*

The characterization of the local extrema in terms of the Frechet derivative is similar to one known from the finite-dimensional calculus.

Theorem 103 *Let* $x_0 \in U$ *be a local maximum (minimum) of the functional* $\phi : U \to R$ *. If functional* ϕ *is Frechet differentiable at* x_0 *then* $D\phi(x_0) = 0$*.*

Proof. For concreteness assume x_0 is a local maximum (the proof is the same if x_0 is a local minimum). Assume that $D\phi(x_0) \neq 0$, then $\exists h \in U$ such that

$$D\phi(x_0)h \neq 0. \tag{4.53}$$

Moreover, we can always choose in such a way that $D\phi(x_0)h > 0$ (if for some $h^* : D\phi(x_0)h^* < 0$ simply define $h = -h^*$). Clearly, $h \neq 0$ and

$$\phi(x_0 + h) - \phi(x_0) = \|h\| \left(D\phi(x_0)\frac{h}{\|h\|} + \frac{1}{\|h\|}o(h)\right). \tag{4.54}$$

Let

$$\alpha = D\phi(x_0)\frac{h}{\|h\|} > 0 \tag{4.55}$$

and

$$h_n = \frac{h}{n}. \tag{4.56}$$

By definition of $o(h)$ there exists $n_0 \in N$ such that for any $n > n_0$

$$\left|\frac{1}{\|h_n\|}o(h_n)\right| < \frac{\alpha}{2}. \tag{4.57}$$

Now take any $\varepsilon > 0$. The definition of h_n implies that there exists $n_1 \in N$ such that for any $n > n_1$

$$\|h_n\| < \varepsilon \tag{4.58}$$

(for example, take any $n_1 > \|h\|/\varepsilon$). Let $n^* = \max(n_0, n_1) + 1$ then $\|(x_0 + h_{n^*}) - x_0\| < \varepsilon$ and

$$\phi(x_0 + h_{n^*}) - \phi(x_0) = \|h\| \left(D\phi(x_0)\frac{h}{\|h\|} + \frac{1}{\|h\|}o(h)\right). \tag{4.59}$$

But according to (4.55), (4.57)

$$\|h\| \left(D\phi(x_0)\frac{h}{\|h\|} + \frac{1}{\|h\|}o(h)\right) \geq \|h\| \left(\alpha - \frac{\alpha}{2}\right) > 0, \tag{4.60}$$

which contradicts the assertion that x_0 is a local maximum. ■

We will finish this section by proving the following lemma.

Lemma 104 *Let mapping $\phi : U \to V$ be continuously Frechet differentiable and $c : [0,1] \to U$ be a mapping continuously differentiable at $(0,1)$ and continuous on $[0,1]$ with $c(0) = x$ and $c(1) = y$. Then*

$$\phi(y) - \phi(x) = \int_0^1 D\phi(c(t)) \cdot c'(t)dt \tag{4.61}$$

Proof. Define $g(t) = (\phi \circ c)(t)$ to be a composite mapping and use the chain rule to get

$$g'(t) = D\phi(c(t)) \cdot c'(t) \tag{4.62}$$

and use Newton-Leibnitz formula. ■

Finally, note that the previous lemma can be written in a differential form.

$$\frac{d\phi(c(t))}{dt} = D\phi(c(t)) \cdot c'(t). \tag{4.63}$$

4.5 Functionals and Gateaux Derivatives

Let V be a normed vector space. Given a functional $L : V \to R$ and an element $a \in V$ with $\|a\| = 1$ define a function $F : R \to R$ by

$$F_{\mathbf{a}}(t) = L(u + at). \tag{4.64}$$

Assume that $F(t)$ is differentiable. Then we can give the following definition:

Definition 105 *The Gateaux directional derivative of functional $L(\cdot)$ in direction a is defined by*

$$\frac{\partial L}{\partial a}(u) = F'_{\mathbf{a}}(0). \tag{4.65}$$

Functionals for which the Gateaux directional derivative exists are called Gateaux differentiable. One might consider Gateaux derivative as a linear functional that maps $a \in V$ into a real number according to (4.65). If we take this point of view we will be able to compare the Gateaux derivative with the Frechet derivative and ask under which conditions do they coincide. The question is important since the Gateaux derivatives is much easier to calculate but the Frechet derivative appears in the first order characterization of local extrema of the functionals. But before we do this, let us consider some examples.

Example 106 *Let $V = R^n$ and $f(\mathbf{x})$ be a differentiable function, then equation (4.65) implies that*

$$\frac{\partial f}{\partial \mathbf{a}} = \nabla f \cdot \mathbf{a}, \tag{4.66}$$

which is a usual definition of a directional derivative.

Example 107 *Let $V = C([0,1])$ be the space of continuous functions with a norm*

$$\|f\| = \max_{x \in [0,1]} |f(x)|. \tag{4.67}$$

Consider a functional

$$L(u) = \int_0^1 u(x)dx. \tag{4.68}$$

Then

$$F_{\mathbf{a}}(t) = \int_0^1 (u(x) + a(x)t)dx = \int_0^1 u(x)dx + t\int_0^1 a(x)dx \tag{4.69}$$

and therefore,

$$\frac{\partial L}{\partial a} = F'_{\mathbf{a}}(0) = \int_0^1 a(x)dx. \tag{4.70}$$

Consider a problem

$$\max_{u \in V} L(u) \tag{4.71}$$

and let u^* solve (4.71). In the previous section we proved that the Frechet derivative $DL(u^*) = 0$. Frechet derivative is, however, usually difficult to calculate directly. Fortunately, the following result holds:

Theorem 108 *Let $L : V \to R$ be is Frechet differentiable at u^*. Then all its Gateaux directional derivatives exist, moreover both types of derivatives viewed as functionals coincide, i. e.*

$$\frac{\partial L(u)}{\partial a} = DF(u)a. \tag{4.72}$$

Proof. Define path $c(\cdot)$ in V by

$$c(t) = u + at, \tag{4.73}$$

then the result follows from Lemma 105. ■

Combining this result with the results from the previous section we arrive at the following theorem.

Theorem 109 *Let functional $L : V \to R$ achieve maximum at $u^* \in V$. If L is Frechet differentiable at u^* then for any $a \in V$*

$$\frac{\partial L}{\partial a} = 0. \tag{4.74}$$

4.6 Euler Equation

Let $V = C^2(\Omega)$ be the space of functions continuous on Ω and twice continuously differentiable on its interior, where $\Omega \subset R^n$ is a compact set with a smooth boundary, Σ, and consider the following functional

$$\int_\Omega L(u, \nabla u, x)dx, \tag{4.75}$$

where $L : R^{2n+1} \to R$ is twice continuously differentiable. We are interested in finding the maxima of the functional (4.75) over V. This problem is called *calculus of variations* problem. We will consider two cases: (i) free boundary, in which no additional constraints are imposed on function $u(\cdot)$ and (ii) fixed boundary, where we assume that $u(x) = u_0(x)$ on Ω for a given function $u_0(\cdot)$. The necessary condition for the maximum are given by the following theorem:

Theorem 110 *Let u^* maximizes functional (4.75) over V. Then it solves the following boundary problem*

$$\sum_{i=1}^{n} \frac{\partial}{\partial x_i}\left(\frac{\partial L}{\partial u_{x_i}}\right) = \frac{\partial L}{\partial u} \tag{4.76}$$

$$\sum_{i=1}^{n}\left(\frac{\partial L}{\partial u_{x_i}} n_i(x)\right) = 0 \text{ on } \Sigma, \tag{4.77}$$

where $\mathbf{n}$ is the unit vector normal to the boundary. If the value of $u(\cdot)$ on the boundary is fixed then boundary condition (4.77) is replaced by

$$u^*(x) = u_0(x) \text{ on } \Sigma. \tag{4.78}$$

Equation (4.76) is known as Euler equation.

Proof. Let us calculate the Gateaux directional derivative of F for some $h \in C^2(\Omega)$. For this purpose, define

$$F_h(t) = \int_\Omega L(u^* + ht, \nabla u^* + \nabla ht, x)dx. \tag{4.79}$$

Since u^* maximizes (4.75) over V, function $F_h(t)$ should achieve maximum at $t = 0$ for any h. Our assumptions on $L(\cdot)$ assure that $F_h(\cdot)$ is differentiable at $t = 0$, therefore $F_h'(0) = 0$. Calculating the derivative one obtains

$$\int_\Omega (\frac{\partial L}{\partial u} h + \sum_{i=1}^{n} \frac{\partial L}{\partial u_{x_i}} h_{x_i}) dx = 0. \tag{4.80}$$

Define vector $\mathbf{a}$ by

$$a_i = \frac{\partial L}{\partial u_{x_i}} \tag{4.81}$$

and note that

$$\sum_{i=1}^{n} a_i h_{x_i} = div(h\mathbf{a}) - \sum_{i=1}^{n} h div(\mathbf{a}). \tag{4.82}$$

Now using the Gauss-Ostrogradsky theorem (Theorem 14) equation (4.80) can be written as

$$\int_\Omega (\frac{\partial L}{\partial u} - \sum_{i=1}^{n} \frac{\partial}{\partial x_i}(\frac{\partial L}{\partial u_{x_i}})) h dx + \oint_\Sigma \mathbf{a} h d\Sigma. \tag{4.83}$$

If the value of u on the boundary is free both the volume and the surface term in (4.83) should vanish for arbitrary $h \in C^2(\Omega)$ and we arrive at system (4.76), (4.77). If, on the other hand, the value of $u(\cdot)$ is fixed on the boundary $h(\cdot)$ should be zero on the boundary for any admissible variation and the surface term in (4.83) vanishes identically. Therefore, the necessary conditions for maximum in that case are (4.76) and (4.78). ∎

4.7 Optimal Control

Let V be the set of piecewise continuously differentiable functions on $[0, 1]$ and consider the problem

$$\max \int_0^1 L(y, u, x) dx \tag{4.84}$$

$$s.t. \frac{dy}{dx} = f(x, y, u). \tag{4.85}$$

Such a problem is called an *optimal control* problem, y is called the state variable, u is called the control variable, and (4.85) is called the evolution equation. The set of admissible controls, U, is assumed to be some subset of a set of piecewise continuously differentiable functions. If one can solve the evolution equation for u in terms of x, y, and y_x and substitute it into the objective function (4.84), then one will arrive at the calculus of variations problem. Otherwise, the solution is given by the following well-known result:

Theorem 111 *Let U be a set of admissible controls and assume a piecewise continuously differentiable function u^* maximizes (4.84) subject to (4.85). Then there exists a differentiable function $\lambda(\cdot)$ (called a costate variable) such that*

$$\frac{d\lambda}{dx} = -\frac{\partial H}{\partial y} \tag{4.86}$$

$$u^*(x) \in \arg\max_{u(\cdot)\in U} H(x,u;\lambda(x),y(x)), \tag{4.87}$$

where $H(x,u;\lambda,y)$, known as the Hamiltonian, is defined by

$$H(x,u;\lambda,y) = L(y,u,x) + \lambda f(x,y,u). \tag{4.88}$$

Moreover, the differential equations (4.85) and (4.86) should be supplemented by boundary conditions, which are

$$y(z) = a \tag{4.89}$$

if the value of y at z is fixed and

$$\lambda(z) = 0 \tag{4.90}$$

if the value y at z is free for $z \in \{0,1\}$. Condition (4.90) is known as the transversality condition.

Let us consider the following example.

Example 112 *Solve the problem*

$$\max \int_0^1 L(y,u,x)dx \tag{4.91}$$

$$s.t. \frac{dy}{dx} = u, \; y(0) = 0, \; y(1) = 1. \tag{4.92}$$

First, form the Hamiltonian

$$H = L(y,u,x) + \lambda u. \tag{4.93}$$

Then according to equation (4.86)

$$\lambda_x = -L_y, \tag{4.94}$$

while equation (4.87) implies that

$$H_u = L_u + \lambda = 0. \tag{4.95}$$

Therefore,

$$L_{ux} = L_y \tag{4.96}$$

or (taking into account $y_x = u$)

$$L_{xy_x} = L_y. \tag{4.97}$$

But this is exactly the Euler equation.

4.8 Examples

Example 113 *Solve:*

$$\max \frac{1}{2} \int_0^{2\pi} (u^2 - u_x^2) dx. \tag{4.98}$$

$$s.t. \; u(0) = u_0 \tag{4.99}$$

The Euler equation for this problem is

$$u_{xx} + u = 0 \tag{4.100}$$

and the transversality condition

$$u_x(2\pi) = 0. \tag{4.101}$$

The general solution to the equation (4.100) is

$$u(x) = C_1 \cos x + C_2 \sin x, \tag{4.102}$$

for arbitrary constants C_1 and C_2. From the transversality condition

$$-C_1 \sin(2\pi) + C_2 \cos(2\pi) = 0 \tag{4.103}$$

we find $C_2 = 0$ and finally, using the initial condition

$$u(x) = u_0 \cos x. \tag{4.104}$$

Example 114 *Let* $\Omega = \{(x, y) : x^2 + y^2 < 1\}$. *Solve the following calculus of variations problem:*

$$\max -\frac{1}{2} \int_{\Omega} (u_x^2 + u_y^2) dx dy \tag{4.105}$$

$$s.t. u(x, y) = x \text{ on } \partial\Omega. \tag{4.106}$$

The Euler equation for this problem is

$$\Delta u = 0. \tag{4.107}$$

The general solution to this problem is given by formula (2.111). Since according to the boundary condition for $r = 1$

$$u(r, \phi) = \cos\phi \tag{4.108}$$

one obtains $D_1 = 0$, $A_1 = 1$, $A_n = 0$ *for* $n \geq 2$ *and* $B_n = 0$. *Therefore,*

$$u(r, \phi) = r\cos\phi \tag{4.109}$$

or

$$u(x, y) = x. \tag{4.110}$$

Example 115 *Solve the following optimal control problem:*

$$\max \int_0^{\tau} (xu - \frac{1}{2}u^2) dt \tag{4.111}$$

$$s.t. \frac{dx}{dt} = x - u, \; x(0) = x_0. \tag{4.112}$$

where $\tau \neq \pi/2 + \pi k$ *for some* $k \in Z$. *Form a Hamiltonian*

$$H = xu - \frac{1}{2}u^2 + \lambda(x - u). \tag{4.113}$$

The first order conditions are

$$\begin{cases} dx/dt = x - u \\ d\lambda/dt = -\partial H/\partial x = -u - \lambda \\ \partial H/\partial u = x - u - \lambda = 0 \\ x(0) = x_0, \; \lambda(\tau) = 0. \end{cases} \tag{4.114}$$

Differentiating the third equation of the system one obtains:

$$x_t - u_t - \lambda_t = 0, \tag{4.115}$$

therefore

$$u_t - x_t = -\lambda_t = u + \lambda = x. \tag{4.116}$$

On the other hand,

$$u_t - x_t = u_t - x + u, \tag{4.117}$$

therefore

$$u_t = 2x - u. \tag{4.118}$$

Consider the system

$$\begin{cases} x_t = x - u \\ u_t = 2x - u \end{cases}. \tag{4.119}$$

Its general solution is given by

$$\begin{cases} x(t) = A\cos t + B\sin t \\ u(t) = (A - B)\cos t + (A + B)\sin t, \end{cases} \tag{4.120}$$

where A and B are arbitrary constants. From the initial condition $A = x_0$. Writing

$$\lambda(t) = x(t) - u(t) = B\cos t - x_0 \sin t \tag{4.121}$$

one can find B from

$$B\cos\tau - x_0\sin\tau = 0, \tag{4.122}$$

provided $\tau \neq \pi/2 + \pi k$ for some $k \in Z$,

$$B = x_0 \tan\tau. \tag{4.123}$$

Therefore,

$$\begin{cases} x(t) = \frac{x_0\cos(t-\tau)}{\cos\tau} \\ u(t) = \frac{x_0(\cos(t-\tau)+\sin(t-\tau))}{\cos\tau} \end{cases}. \tag{4.124}$$

4.9 Exercises

1. Show that expressions (4.1), (4.2), (4.7) indeed define norms on the respective vector spaces. Why (4.7) does not define a norm on the set of *functions* whose p^{th} power is Lebesgue integrable?
2. Prove that any converging sequence is fundamental.
3. Show that R^n with the Euclidean norm and $C(\Omega)$ are complete normed spaces. Assume that Ω is closed. Show that then $L^p(\Omega)$ is complete.
4. Let $X = R$ and consider a function

$$d(x, y) = \frac{|x - y|}{1 + |x - y|}. \tag{4.125}$$

Show that $d(\cdot,\cdot)$ is a metric but there exists no norm $\|\cdot\|$ on R such that

$$d(x, y) = \|x - y\|. \tag{4.126}$$

(Hint: note that in general $d(\lambda x, 0) \neq \lambda d(x, 0)$).

5. Let (X,d) be a metric space. Call subset $F \subset X$ closed if it contains the limits of all its converging sequences. Prove that if (X,d) is complete, so is (F,d).
6. Show that equation (4.23) defines a norm on a Euclidean space.
7. Prove that an open ball is an open set and a closed ball is a closed set.
8. Let (X,d_X) and (Y,d_Y) be metric spaces and $f: X \to Y$ is a function. Define what does it mean for f to be continuous. (Hint: use the usual definition for continuity of a function from R^m into R^n and replace the Euclidean distances with the the respective metrics).
9. Let $\Omega = \{(x,y) : x^2+y^2 < 1\}$. Solve the following calculus of variations problem:

$$\max -\frac{1}{2}\int_{\Omega}(u_x^2+u_y^2-u^2)dxdy \tag{4.127}$$

$$s.t.u(x,y) = x \text{ on } \partial\Omega. \tag{4.128}$$

10. Solve the following optimal control problem:

$$\max \int_0^T (u - \frac{u^2}{2})dt \tag{4.129}$$

$$s.t.\frac{dx}{dt} = 2x - u, \ x(0) = x_0. \tag{4.130}$$

Consider the limit of the solution as $T \to \infty$.
11. Consider the following calculus of variations problem:

$$\min \int_0^1 L(u,u_x)dx, \tag{4.131}$$

where $L(u,v)$ is convex in v. Define the Legandre transform of $L(\cdot,\cdot)$ by

$$H(p,u) - \max_v(pv - L(u,v)). \tag{4.132}$$

and denote the maximizer by $p(u,v)$. Now suppose that u_x satisfies Euler equation and $p = p(u,u_x)$. Prove that the following system holds:

$$\begin{cases} u_x = \partial H/\partial p \\ p_x = -\partial H/\partial u \end{cases}. \tag{4.133}$$

4.10 Bibliographic Notes

A very good modern approach that covers the material in the first part of this Chapter: Banach and Polish spaces, Frechet and Gateaux derivatives, and the geometrical interpretation of the derivative is Abraham, Mardsen, and Ratiu (1988). This tools are important not only for the development of the optimal control theory and the calculus of variations, but also for formulating the screening model with an infinite-dimensional type. This case, though probably the most relevant from the economic point of view, is not yet studied in the literature.

A good text on the optimal control is Lewis (1986), which also covers some numerical techniques for solving optimal control problems. Note that traditional books on optimal control treat such problems in one dimension. The control variables are assumed to be governed by a system of ordinary differential equations. Some work on the optimal control when the control variables is governed by the systems of partial differential equations was performed by Lions (1971). However, since he considered systems of the partial differential equations of the second order that usually govern the physical systems, his results cannot be applied directly to the screening problems. I develop some multidimensional extensions in Chapters 7 and 8 of this book when I will discuss screening problems. An excellent introduction to the calculus of variations is given in Gelfand and Fomin (1963). For a more advanced treatment, see Eberhard (1984).

5 Miscellaneous Techniques

In the previous chapters I reviewed the main mathematical techniques that are used to solve screening models. This chapter reviews techniques, which may be sometimes useful in screening applications, but whose use is not that prevalent as the use of the techniques collected in the first four chapters. I collect all of them in this chapter, though they belong to different areas of mathematics. Therefore, presentation in this chapter is necessarily more eclectic then in the rest of the book.

5.1 Distributions as Solutions of Differential Equations

Sometimes in applications we have to deal with probability distributions with atoms or singular probability distributions. Such distributions may arise endogenously even if all the data of the problem are characterized by distributions that possess a density function. For example, as we will see below, the Lagrange multiplier on the participation constraint could be a probability distribution that assigns a positive mass to certain subsets of type space with Lebesgue measure zero. The classical density for such distributions does not exist, which seems to create a difficulty for the application of the optimal control technique, which is based on solving a system of partial differential equations. A way to deal with this problem is to extend the notion of a function to cover such objects as densities of discrete and singular random variables and extend the notion of a solution of a (partial) differential equation accordingly. That leads us to the concepts of a distribution and a generalized solution of a differential equation. Another way to deal with this problems, which uses measure-theoretic characterization is presented in Tikhomirov and Ioffe (1979). Using distribution theoretic technic, however, makes the characterization simpler. Therefore, I will start with a brief review of the theory of distributions. For a detailed exposition, see Vladimirov (2002).

5.1.1 A Motivating Example

Let $\Omega = R^2/\{\mathbf{0}\}$ and consider the following Dirichlet problem, formulated in the polar coordinates

$$\Delta u = 2\pi\rho(r), \tag{5.1}$$
$$u(1) = 0. \tag{5.2}$$

Assume that

$$\rho(r;\sigma^2) = \frac{1}{2\pi\sigma^2}\exp(-\frac{r^2}{2\sigma^2}). \tag{5.3}$$

Since both equation (5.1) and the boundary condition do not depend on ϕ the solution should be rotationally invariant solution, that is it should depend only on r. Therefore, $u(\mathbf{r}) = u(r)$, where $u(\cdot)$ should the following ordinary differential equation

$$\frac{1}{r}\frac{d}{dr}(ru'(r)) = 2\pi\rho(r). \tag{5.4}$$

One can verify by a direct substitution that

$$u(r;\sigma^2) = C\ln r + \int_r^1 \frac{\exp(-\frac{x^2}{2\sigma^2})}{x}dx, \tag{5.5}$$

where C is an arbitrary constant solves (5.1)-(5.2). Now, if one takes the limit of (5.5) as σ^2 goes to zero she will obtain

$$u(r) = C\ln r. \tag{5.6}$$

A legitimate question is: How does the corresponding function $\rho(r,0)$ look like? Let us start with studying the properties of $\rho(r;\sigma^2)$ for $\sigma^2 > 0$. Note that $\rho(0;\sigma^2) = (2\pi\sigma^2)^{-1}$ and for any $R > 0$

$$\iint_{\{(r,\phi:r\leq R\}} \rho(r;\sigma^2)drd\phi = 1 - \exp(-\frac{R^2}{2\sigma^2}). \tag{5.7}$$

Therefore,

$$\begin{cases} \lim_{\sigma^2 \to 0} \rho(r;\sigma^2) = 0 \text{ for } r \neq 0 \\ \lim_{\sigma^2 \to 0} \rho(0;\sigma^2) = \infty \\ \lim_{\sigma^2 \to 0} \iint_{\{(r,\phi: r \leq R\}} \rho(r;\sigma^2) dr d\phi = 1. \end{cases} \tag{5.8}$$

Therefore "function" $\rho(r,0)$ (let us call it $\delta(r)$) should have the following properties: it should be zero everywhere expect for the origin, where it is infinite and its integral over any set containing the origin is one. Therefore, $u(r) = \ln r$ solves

$$\begin{cases} \Delta u = 2\pi\delta(r) \\ u(1,\phi) = 0 \end{cases} . \tag{5.9}$$

Intuitively, $\delta(r)$ is the density of a random variable, which takes value zero with probability one. The problem is that a function with described properties does not exist. Indeed, if the function is zero everywhere expect one point then its integral over any set should be equal to zero. However, it turns out that it is possible to define an object that will have all the properties of $\delta(r)$. Such an object is called a distribution or a generalized function.

5.1.2 The Set of Test Functions and Its Dual

Let us consider the set of bounded continuous functions on R^n, $C(R^n)$ and endow it with a norm

$$\|f\| = \sup_{x \in R^n} |f(\mathbf{x})| . \tag{5.10}$$

Definition 116 *We will say that a sequence of bounded continuous functions $\{f_n\}_{n=0}^{\infty}$ converges to a bounded continuous function f uniformly on R^n and write*

$$f_n \rightrightarrows f \tag{5.11}$$

if $\forall \varepsilon > 0$ there exists $n_0 \in N$ such that

$$\|f - f_n\| < \varepsilon \tag{5.12}$$

for any $n > n_0$.

Now let $C_0^\infty(R^n)$ be the set of infinitely differentiable functions from R^n into R with a compact support, i. e. the functions that vanish outside some compact set $K \subset R^n$. Let $h^{(k)}$ denote the kth derivative of h.

Definition 117 *We will say that a sequence of functions $\{h_n\}_{n=0}^{\infty}$, $h_n \in C_0^\infty(R^n)$ converges to a function $h \in C_0^\infty(R^n)$ and write*

$$h_n \to h \tag{5.13}$$

if

$$h_n^{(k)} \rightrightarrows h^{(k)} \tag{5.14}$$

for any k.

Our next task is to define linear continuous functionals on . The definition of linearity is straightforward.

Definition 118 *A functional* $\varphi : C_0^\infty(R^n) \to R$ *is called linear if* $\forall\ h_1$, $h_2 \in C_0^\infty(R^n)$ *and* $\forall \alpha, \beta \in R$:

$$\varphi(\alpha h_1 + \beta h_2) = \alpha\varphi(h_1) + \beta\varphi(h_2). \tag{5.15}$$

To define the notion of continuity we will use the above given definition of convergence in $C_0^\infty(R^n)$.

Definition 119 *A functional* $\varphi : C_0^\infty(R^n) \to R$ *is called continuous if for any sequence* $\{\ h_n\}_{n=0}^\infty$, $h_n \in C_0^\infty(R^n)$ *such that*

$$h_n \to h \tag{5.16}$$

the numerical sequence $\varphi(\ h_n)$ *converges to* $\varphi(\ h)$.

Below we will call set $C_0^\infty(R^n)-$ the set of the test functions. Define its dual $C_0^\infty(R^n)^*$ by

$$D = C_0^\infty(R^n)^* = \{\varphi : \varphi \text{ is linear and continuous functional}\}. \tag{5.17}$$

Elements of set D are called *distributions or generalized functions.* D is obviously a linear space.

Definition 120 *Elements of D are called distributions. Distribution* $\phi \in D$ *is called positive if* $\phi(\ h) \geq 0$ *for any* $h \in C_0^\infty(R^n)$ *such that* $h(\mathbf{x}) \geq 0$ *for all* $\mathbf{x} \in R^n$.

5.1.3 *Examples of Distributions*

Our first example illustrates that the set of distributions is a natural extension of the set of measurable functions in a sense that there exists a natural embedding of the latter into the former.

Example 121 *Let* $f : R^n \to R$ *be a measurable function. Define functional* φ_f *on* $C_0^\infty(R^n)$ *by*

$$\varphi_f(h) = \int\limits_{R^n} f(\mathbf{x})h(\mathbf{x})\mathbf{dx}. \tag{5.18}$$

Obviously, $\varphi_f \in D$. *Therefore, there exists a natural embedding of the set of measurable functions (and in fact, Lebesgue integrable functions) into D.*

This example can be generalized in a following way. Let μ be any measure (recall that a measure on the $\sigma-$algebra $B(R^n)$ of the Borel subsets of R^n is a nonnegative countably additive function $\mu : B(R^n) \to R$). Then

$$\varphi_f(h) = \int_{R^n} h(\mathbf{x}) d\mu(\mathbf{x}) \tag{5.19}$$

is a distribution. Therefore, measures can be embedded in the set of distributions.

The second example intends to show that some distributions cannot be naturally identified with any function. Therefore, the set of distributions contains some genuinely new objects.

Example 122 *Define functional $\varphi_{s,\mathbf{a}}$ on $C_0^\infty(R^n)$ in the following way*

$$\varphi_{s,\mathbf{a}}(h) = h(\mathbf{a}) \tag{5.20}$$

for some $\mathbf{a} \in R^n$, i. e. the functional maps each function into its value at some fixed point. It is straightforward (and is left as an exercise to the reader) to check that $\varphi_{s,\mathbf{a}} \in D$ and there exists no Lebesgue integrable function $f : R^n \to R$ such that

$$\varphi_{s,\mathbf{a}}(h) = \int_{R^n} f(\mathbf{x}) h(\mathbf{x}) \mathbf{dx}. \tag{5.21}$$

The exists, however, a measure μ such that $\varphi_{s,\mathbf{a}}$ can be represented as (5.19). Indeed, for any $A \subset R^n$ set $\mu(A) = 1$ if $\mathbf{a} \in A$ and $\mu(A) = 0$ otherwise. Note that μ is a measure, it is defined for all (not only Borel) subsets of R^n and (5.19) holds. Let us call distributions that can be represented in the form (5.21) regular. The exists a natural isomorphism between the set of the regular distributions and the set of the measurable functions. We will call such distributions regular. Otherwise, a distribution is called singular.

Let $f(\cdot)$ be defined by

$$f(x; a, \sigma^2) = \frac{1}{2\pi\sigma^2} \exp(-\frac{(x-a)^2}{2\sigma^2}). \tag{5.22}$$

and assume for simplicity $n = 1$. Then, using the Taylor expansion of $h(\cdot)$ up to the second order terms

$$\int_{-\infty}^{\infty} h(x) f(x; a, \sigma^2) dx = h(a) + \frac{1}{2} \int_{-\infty}^{\infty} h''(\zeta)(x-a)^2 f(x; a, \sigma^2) dx, \tag{5.23}$$

where $\zeta \in (x, a)$. According to the Mean Value Theorem (see, for example, Grossman, 1992):

$$\int_{-\infty}^{\infty} h(x) f(x; a, \sigma^2) dx = h(a) + \frac{1}{2} h''(\xi)\sigma^2 \tag{5.24}$$

for some $\xi \in (x, a)$. Therefore,

$$\lim_{\sigma \to 0} \int_{-\infty}^{\infty} h(x) f(x; a, \sigma^2) dx = h(a). \tag{5.25}$$

Recall, that I introduced symbol $\delta(x)$ as the limit of $f(x; 0, \sigma^2)$. However, we were unable to give the precise sense to the symbol and the meaning of the limit. Now we can do it. With each $f(x; a, \sigma^2)$ one can associate an element of D according to

$$\varphi_{\sigma, \mathbf{a}}(h) = \int_{-\infty}^{\infty} h(x) f(x; a, \sigma^2) dx. \tag{5.26}$$

Then one can write

$$\varphi_{s, \mathbf{a}}(h) = \lim_{\sigma \to 0} \varphi_{\sigma, \mathbf{a}}(h) \tag{5.27}$$

for any test function $h \in C_0^\infty(R^n)$ or, for short,

$$\varphi_{s, \mathbf{a}} = \lim_{\sigma \to 0} \varphi_{\sigma, \mathbf{a}}. \tag{5.28}$$

If one symbolically writes

$$\varphi_{s, \mathbf{a}}(h)) = \int_{R^n} \delta(\mathbf{x} - \mathbf{a}) h(\mathbf{x}) \mathbf{dx}, \tag{5.29}$$

then

$$\delta((\mathbf{x} - \mathbf{a}) = \lim_{\sigma \to 0} f(x; a, \sigma^2). \tag{5.30}$$

Note that (5.29) should be considered as the definition of the symbol $\delta(\mathbf{x} - \mathbf{a})$, known as Dirac's delta function. Integral in (5.29) is again purely symbolic and is defined to be equal to the left hand side. This symbolism, however, turns out to be very useful.

The above examples illustrate that not all distributions can be identified with functions. We will, however, often use notation $\phi(\mathbf{x})$ for arbitrary distribution ϕ. To fill this notation with some meaning let us give the following definition.

Definition 123 *Let $\phi \in D$ and $A \subset R^n$ is a Borel set. Then ϕ is called regular on A if $\phi(g)$ can be represented in the form (5.19) with some measurable function $h(\cdot)$ for any $g \in C_0^\infty(R^n)$ whose support is contained in A. For any $\mathbf{x} \in A$ define $\phi(\mathbf{x}) = h(\mathbf{x})$.*

Therefore, notation $\phi(\mathbf{x})$ can be given literal meaning at any point at which ϕ is regular. For example, $\delta(\mathbf{x}) = 0$ for $\mathbf{x} \neq \mathbf{0}$.

5.1.4 The Derivative of a Distribution

Let g be a continuously differentiable function and consider the following functional on $C_0^\infty(R^n)$:

$$\psi_g(h)) = \int_{-\infty}^{\infty} g(x)h(x)dx. \tag{5.31}$$

(for simplicity, I assumed $n = 1$). It will be natural to define a derivative of this functional by

$$\psi_g'(h) = \int_{-\infty}^{\infty} g'(x)h(x)dx. \tag{5.32}$$

Our aim is to extend this definition of the derivative from functionals that can be represented in form (5.31) to the whole D. For this purpose integrate equation (5.32) by parts. Then using the fact that $h(\cdot)$ has a compact support, one obtains:

$$\psi_g'(h) = -\int_{-\infty}^{\infty} g(x)h'(x)dx = -\psi_g(h'). \tag{5.33}$$

We will use this result to define derivatives of an arbitrary functional from D.

Definition 124 *For any $\varphi \in D$ its derivative is an element $\varphi' \in D$ defined by*

$$\varphi'(h)) = -\varphi(h') \tag{5.34}$$

for $\forall\ h \in C_0^\infty(R^n)$.

Note that all distributions are differentiable in the sense of formula (5.34). Since any Lebesgue integrable function can be naturally embedding in the set of distributions any such function is differentiable in the sense of distributions. This observation will allow us to define solution for partial differential equations, which are not differentiable in the classical sense.

Example 125 *Define the Heviside step function, $\theta(\cdot)$, by*

$$\theta(x) = \begin{cases} 1, & \text{for } x \geq 0 \\ 0, & \text{for } x < 0 \end{cases} \tag{5.35}$$

and let distribution φ be defined as

$$\varphi(h)) = \int_{-\infty}^{\infty} h(x)\theta(x)dx = \int_{0}^{\infty} h(x)dx. \tag{5.36}$$

With some abuse of terminology we can identify distribution $\varphi(\cdot)$ with function $\theta(\cdot)$ (recall the natural embedding of the set of Lebesgue integrable functions into the set of distributions). The derivative of distribution $\varphi(\cdot)$ is defined by (5.34) and can be calculated as

$$\varphi'(h) = -\int_0^\infty h'(x)dx = h(0) = \int_{-\infty}^\infty h(x)\delta(x)dx. \tag{5.37}$$

The last equality uses the symbolic definition of the Dirac's delta-function. The result of this example can be summarized as:

$$\theta'(x) = \delta(x). \tag{5.38}$$

In general, if a Lebesgue integrable function $f(\cdot)$ has a jump at point a and both left and right limits at this point exists

$$f'(a) = (f(a+0) - f(a-0))\delta(x-a). \tag{5.39}$$

5.1.5 The Product of a Distribution and a Test Function and the Product of Distributions

Let $h \in C_0^\infty(R^n)$ be a test function and $\phi \in D$ a distribution.

Definition 126 *We will say that $\varphi \in D$ is a product of test function h and distribution ϕ and write $\varphi = h\phi$ if for any $f \in C_0^\infty(R^n)$*

$$\varphi(f) = \phi(hf). \tag{5.40}$$

Note that if ϕ can be represented in form (5.31) with some measurable g then φ can be represented in the same form with $g^* = g\ h$. Therefore, the definition seems quite natural. To see how the definition works, let us consider the following example.

Example 127 *Let $h \in C_0^\infty(R^n)$ and let us find the following product*

$$h(x)\delta(x-a). \tag{5.41}$$

For any $f \in C_0^\infty(R^n)$

$$\delta(x-a)(h(x)f(x)) = h(a)f(a) = h(a)\delta(x-a)(f(x)), \tag{5.42}$$

therefore

$$h(x)\delta(x-a) = h(a)\delta(x-a). \tag{5.43}$$

In particular,

$$(h(x) - h(a))\delta(x-a) = 0. \tag{5.44}$$

Using this fact and taking $h(x) = x$ it can be shown that the general solution to the equation

$$(x-a)\varphi(x) = 0 \tag{5.45}$$

in the set of distributions is

$$\varphi(x) = C\delta(x-a), \tag{5.46}$$

where C is an arbitrary constant.

With a little bit more of technical work one can define a product of two distributions, provided at least one of them is regular. It is, however, impossible to extend this definition on the case, when both distribution are singular. I will not discuss the arising technical difficulties here. An interested reader could consult Vladimirov (2002).

5.1.6 The Resultant of a Distribution and a Dilation

The notions of the *resultant* and the *dilation operator* were first introduced by Meyer (1966). He defined them on the space of measures only. Here I am going to extend these definitions for arbitrary distributions. Let $\phi \in D$ is a distribution and $K \subset C_0^\infty(R^n)$.

Definition 128 *A point $\mathbf{r} \in R^n$ is called the resultant of distribution ϕ with respect to K if*

$$\phi(h) = h(\mathbf{r}) \tag{5.47}$$

for any $h \in K$. If $\mathbf{a} \in R^n$ is the resultant of ϕ we write

$$r_K(\phi) = \mathbf{r}. \tag{5.48}$$

Let us start with considering the following example.

Example 129 *Let functional $\varphi_{s,\mathbf{a}}$ be defined by (5.20). It is straightforward to check that*

$$r_K(\varphi_{s,\mathbf{a}}) = \mathbf{a}. \tag{5.49}$$

for any $K \subset C_0^\infty(R^n)$.

The resultant need not always exists. Let us consider the following example.

Example 130 *Let $n = 1$ and define $\phi \in D$ by*

$$\phi(h) = \int_0^1 h(x)dx. \tag{5.50}$$

Let $K = C_0^\infty(R^n)$ and let $h_1,\ h_2 \in C_0^\infty(R^n)$ be such that

$$h_1(x) = \begin{cases} x, \ for \ [0,1] \\ 0, for \ |x| > 2 \end{cases}, \quad (5.51)$$

$$h_2(x) = \begin{cases} x^2, \ for \ [0,2] \\ 0, for \ |x| > 3 \end{cases}, \quad (5.52)$$

and $h_i(x) \le 0$ for $x \le 0$. Then

$$\phi(h_1) = \frac{1}{2},\ \phi(h_2) = \frac{1}{3}. \quad (5.53)$$

Suppose the resultant exists. Then

$$h_1(r_K(\phi)) = \frac{1}{2} \Rightarrow r_K(\phi) = \frac{1}{2} \ or \ r_K(\phi) \in (1,2) \quad (5.54)$$

$$h_2(r_K(\phi)) = \frac{1}{3} \Rightarrow r_K(\phi) = \frac{1}{\sqrt{3}} \ or \ r_K(\phi) \in (2,3). \quad (5.55)$$

These conditions are, however, inconsistent. Therefore, the resultant does not exist.

Let $T : D_c \to D_c$ be a linear operator.

Definition 131 *Operator $T : D_c \to D_c$ is called a dilation with respect to the set K if*

$$r_K(T\varphi_{s,\mathbf{a}}) = \mathbf{a}, \quad (5.56)$$

where $\varphi_{s,\mathbf{a}}$ is defined by (5.20).

If K contains all constants and operator T is positive then one can say that T spreads the initial probability measure given by a unit mass at point $\mathbf{a}$ away from that point in such a way that the expectation of all functions in K remains $h(\mathbf{a})$.

Example 132 *Let $\Omega = R^n$ and $K \subset C(R^n)$ consists of all constant functions. Let us define operator T in the following way. Take any probability density function $f \in C(R^n)$ and for any $\mathbf{x}, \mathbf{y} \in R^n$ define*

$$g_{\mathbf{x}}(y) = f(\mathbf{x} - \mathbf{y}). \quad (5.57)$$

Note that $\phi(g_{\mathbf{x}})$ is a continuous function of $\mathbf{x}$ for any $\phi \in D$ and define operator T by

$$T\phi(h)) = \int_\Omega h(\mathbf{x})\phi(g_{\mathbf{x}})\mathbf{dx} \quad (5.58)$$

for any $h \in C(R^n)$. Let us show that T is a dilation with respect to K. Indeed, for any $h \in K$

$$T\varphi_{s,\mathbf{a}}(h)) = \int_\Omega h(\mathbf{x})f(\mathbf{x} - \mathbf{a})\mathbf{dx} = h(\mathbf{a}). \quad (5.59)$$

The last equality follows from the fact that $h(\cdot)$ is constant and $f(\cdot)$ is a probability distribution.

5.1.7 Adjoint Linear Differential Operators and Generalized Solutions of the Partial Differential Equations

Let L be a linear differential operator on the set $C^{\infty}(R^n)$, i. e. $L : C^{\infty}(R^n) \to C^{\infty}(R^n)$ and for $\forall f \in C^{\infty}(R^n)$

$$Lf = \sum_{i=0}^{m} \sum_{j_1 + \dots j_n = i} c^i_{j_1,\dots,j_n}(x) \frac{\partial^i f}{\partial x_1^{j_1} \dots \partial x_n^{j_n}}. \tag{5.60}$$

For any $f \in C^{\infty}(R^n)$ and for any $h \in C_0^{\infty}(R^n)$ define the inner product

$$(f, h) = \int_{R^n} f(x)h(x)dx. \tag{5.61}$$

Definition 133 *Linear differential operator $L^* : C_0^{\infty}(R^n) \to C_0^{\infty}(R^n)$ is called adjoint to L if for $\forall f \in C^{\infty}(R^n)$ and for $\forall\ h \in C_0^{\infty}(R^n)$*

$$(Lf, h) = (f, L^* h). \tag{5.62}$$

This definition generalizes a well-known definition in linear algebra.

Example 134 *Let $n = 1$ and $Lf = f'(x)$. Since for $\forall\ h \in C_0^{\infty}(R^n)$*

$$\int_{-\infty}^{\infty} f'(x)h(x)dx = -\int_{-\infty}^{\infty} f(x)h'(x)dx, \tag{5.63}$$

we find that $L^*\ h = -h'(x)$. In a symbolic form one can write

$$(d/dx)^* = -(d/dx). \tag{5.64}$$

Now recall that for any $f \in C^{\infty}(R^n)$ equation (5.61) defines a linear continuous functional on $C_0^{\infty}(R^n)$, i. e. a distribution. To simplify further notation let us agree to write

$$\phi(h) = (\phi, h) \tag{5.65}$$

for any $\phi \in D$. Consider a linear PDE

$$Lu = \varphi(x), \tag{5.66}$$

where $\varphi(\cdot)$ is in general a distribution. If it is a Lebesgue integrable function we can always think of it as a distribution in the sense of the natural embedding.

Definition 135 *A distribution u is called a generalized solution of equation (5.66) if for $\forall h \in C_0^\infty(R^n)$*

$$(u, L^* h)) = (c, h). \tag{5.67}$$

Let us consider the following example.

Example 136 *Consider an ordinary differential equation*

$$u' = 2\theta(x) - 1 \tag{5.68}$$

where $\theta(\cdot)$ is defined by (5.35). Let us prove that a distribution

$$u(x) = |x| \tag{5.69}$$

solves (5.68). Let $h \in C_0^\infty(R^n)$, then

$$(u, -h'(x)) = -\int_{-\infty}^{\infty} |x|\, h'(x)dx = \int_{0}^{\infty} h(x)dx - \int_{-\infty}^{0} h(x)dx. \tag{5.70}$$

(I used equation (5.64), replaced $|x|$ with x for $x \geq 0$ and $-x$ for $x < 0$, integrated by parts and used the fact that $h(\cdot)$ has a compact support). On the other hand

$$(2\theta(x) - 1, h) = \int_{-\infty}^{\infty} (2\theta(x) - 1)h(x)dx = \int_{0}^{\infty} h(x)dx - \int_{-\infty}^{0} h(x)dx. \tag{5.71}$$

Therefore,

$$(|x|\,, -h') = (2\theta(x) - 1, h), \tag{5.72}$$

which proves our assertion.

5.2 Sobolev Spaces and Poincare Theorem

Let $\Omega \subset R^n$ be an open set. Recall that a Lebesgue measurable function is said to be of class $L^p(\Omega)$ for some $p \geq 1$ if the following integral converges

$$\int_{\Omega} |f(\mathbf{x})|^p \, d\mathbf{x}. \tag{5.73}$$

A standard norm on L^p is defined by

$$\|f\|_{L^p} = (\int_{\Omega} |f(\mathbf{x})|^p \, d\mathbf{x})^{1/p}. \tag{5.74}$$

Sobolev space $H^{m,p}$ is defined in the following way:

Definition 137 *$H^{m,p}(\Omega)$ is the set of all functions f such that f is m times differentiable almost everywhere on Ω with respect to the Lebesgue measure and all its derivatives up to the order of m belong to L^p.*

This space is usually endowed with a norm

$$\|f\|_{m,p} = (\sum_{i=0}^{m} \sum_{j_1+...j_n=i} \int_{\Omega} \left| \frac{\partial^i f(\mathbf{x})}{\partial x_1^{j_1}...\partial x_n^{j_n}} \right|^p \mathbf{dx})^{1/p}. \tag{5.75}$$

The fundamental fact about Sobolev spaces endowed with norm (5.75) is given by the following lemma.

Lemma 138 *$H^{m,p}(\Omega)$ is a Banach space.*

I will not give a proof of this fact here. An interested reader is referred, for example, to Barros-Neto (1973).

Form now on let Ω be bounded. The most important of the Sobolev's spaces from the point of view of screening applications is $H^{1,2}(\Omega)$ which I will denote $H^1(\Omega)$ for short, and $H^{m,1}(\Omega)$. In the latter case one can extend the notion of the Sobolev space to cover distributions. For this purpose one has first define the notion of the integral of a distribution. Let $\Omega \subset R^n$ be a bounded open set and χ_Ω be its characteristic function. Denote by Ω_ε the $\varepsilon-$neighborhood of set Ω. Then the exists function $h_\varepsilon \in C_0^\infty(R^n)$ such that $h_\varepsilon(\mathbf{x}) = 1$ for all $\mathbf{x} \in \Omega$ and $h_\varepsilon(\mathbf{x}) = 0$ for $\mathbf{x} \in R^n \backslash \Omega_\varepsilon$. Now for any distribution ϕ define

$$\int_\Omega \phi(\mathbf{x})d\mathbf{x} = \lim_{\varepsilon \to 0} \phi(h_\varepsilon). \tag{5.76}$$

For example, if $\mathbf{a} \in \Omega$ then

$$\int_\Omega \delta(\mathbf{x} - \mathbf{a})\mathbf{dx} = \delta(h_0) = h_0(\mathbf{a}) = 1 \tag{5.77}$$

in accordance we our previous purely symbolic notation. Now one can define space $H^{m,1}(\Omega)$ for the distributions verbatim the same way it was defined for the functions. Below, I will always assume that $H^{m,1}(\Omega)$ is the set of distributions.

Let $mes(\Omega)$ denote the Lebesgue measure of set Ω. For an arbitrary $s \in H^1(\Omega)$ define its mean value over Ω by. The following fact, known as Poincare lemma is important for some existence proofs in screening literature:

$$\underline{s} = \frac{1}{mes(\Omega)} \int_\Omega s(\alpha)d\alpha. \tag{5.78}$$

Lemma 139 *There exists a constant $M(\Omega)$ such that for all $s \in H^1(\Omega)$,*

$$\|s - \underline{s}\|_{L^2} \leq M(\Omega) |\nabla s|_{L^2}. \tag{5.79}$$

Again, we are not proving this lemma and refer an interested reader to Kinderlehrer and Stampacchia (1980).

5.3 Sweeping Operators and Balayage of Measures

Recall that a measure on the $\sigma-$algebra $B(R^n)$ of the Borel subsets of R^n is a nonnegative countably additive function $\mu : B(R^n) \to R$. It is often convenient to define a signed measure, also known as a charge, by dropping the requirement that μ is non-negative. Let us define the space CH to be the linear space of all charges on $B(R^n)$. With any charge $\nu \in CH$ one can associate a distribution $\phi \in D$ by

$$\phi(h) = \int_{R^n} h(\mathbf{x}) d\nu(\mathbf{x}). \tag{5.80}$$

Therefore, there exists a natural embedding of CH in D and with some abuse of notation one can write $C\,H \subset D$, moreover CH is a linear subspace of D. Therefore the resultants for the charges and dilations from CH into CH are well-defined. I will denote by $M \subset CH$ the subset of measures on $B(R^n)$ and call a dilation operator on CH restricted to M a *sweeping operator.*

Let us define a partial order relation on the set of measures known as *balayage.* Balayage is a French term introduced by Meyer in his (1966) book, which translates into English as *sweeping out.* To define the balayge we have to start with a *convex cone* of function on R^n. Let us give the following definition.

Definition 140 *A subset K of a linear space is called a cone if $\alpha\mathbf{x} \in K$ for any $\mathbf{x} \in K$ and any $\alpha \geq 0$.*

We will say that K is a convex cone if it is convex and it is a cone. Let $\Omega \subset B(R^n)$ and $K \subset C(\Omega)$ is a convex cone that contains all constant functions.

Definition 141 *For λ, $\mu \in M$ define a partial order $\overset{K}{\prec}$ induced by K in the following way*

$$(\lambda \overset{K}{\prec} \mu) \Leftrightarrow (\int_{\Omega} h d\lambda \geq \int_{\Omega} h d\mu) \tag{5.81}$$

for any $h \in K$. If $\lambda \overset{K}{\prec} \mu$ we say that λ is balayage of μ relative to K.

Let $\lambda \overset{K}{\prec} \mu$. An important result is the so called Cartier's theorem that states that λ is balayage of μ relative to K if and only if there exists a sweeping operator $T : M \to M$ such that $\lambda = T\mu$, i. e. the following theorem holds:

Theorem 142 *(Cartier) Let $\Omega \subset R^n$ be a compact set and $K \subset C(\Omega)$ is a convex cone that contains all constant functions. Then $\lambda \overset{K}{\prec} \mu$ if and only if there exists a sweeping operator such that $\lambda = T\mu$.*

I will not prove this theorem here. An interested reader is referred to Meyer (1966). Let us concentrate on a special case, where K is the set of continuous convex functions on Ω. Let us start with the following definitions:

Definition 143 *A measure μ defined on the Borel subsets of Ω is called a probability measure if $\mu(\Omega) = 1$. The set of all probability measures on Ω is denoted $P(\Omega)$.*

Definition 144 *Operator $T : R \times P(\Omega) \to P(\Omega)$ such that $T(t, \mu)$ is measurable in t for $\forall \mu \in P(\Omega)$ is called a transition probability.*

Now we are ready to formulate the following proposition

Proposition 145 *Let K be the convex cone of convex continuous functions on Ω and $T : M \to M$ be a sweeping operator with respect to K . Let $\lambda, \mu \in M$ be two such measures that*

$$\lambda = T\mu. \tag{5.82}$$

Then

$$\mu(\Omega) = \lambda(\Omega) \tag{5.83}$$

$$\int_{\Omega} \mathbf{x} d\mu(\mathbf{x}) = \int_{\Omega} \mathbf{x} d\lambda(\mathbf{x}). \tag{5.84}$$

Proof. Since T is a sweeping operator, the Cartier's Theorem implies that $\lambda \overset{K}{\prec} \mu$, i. e.

$$\int_{\Omega} h d\lambda \geq \int_{\Omega} h d\mu \tag{5.85}$$

for any $h \in K$. Setting $h(\mathbf{x}) = 1$ proves (5.83). Noting that both $h(\mathbf{x}) = \mathbf{x}$ and $h(\mathbf{x}) = -\mathbf{x}$ belong to K establishes (5.84). ■

Note that property (5.83) will hold for any convex cone that contains positive constants, while property (5.84) in general will not hold. Another important convex cone of functions for which both these properties hold is the set of functions convex along rays. We will see in the next part that both these cones arise as classes of implementable surpluses in some screening models.

5.4 Coercive Functionals

Suppose $f : R^m \to R$ is a concave continuous function such that

$$\lim_{\|\mathbf{x}\|\to\infty} f(\mathbf{x}) = -\infty, \tag{5.86}$$

and $K \subset R^m$ is a closed convex set. Then $f(\cdot)$ achieves a maximum at K. This idea can be generalized to the functionals defined on closed convex subsets of $H^1(\Omega)$, where Ω is a convex, open, bounded subset of R^m. We will start by defining the notions of a concave, coercive functional. Since both definitions do not rely on any particular structure in $H^1(\Omega)$ we will give the definition for arbitrary Banach spaces.

Definition 146 *Let X be a Banach space. A functional $h : X \to R$ is called concave if for any $x_1, x_2 \in X$ and any $\lambda \in [0,1]$*

$$h(\lambda x_1 + (1-\lambda)x_2) \geq \lambda h(x_1) + (1-\lambda)h(x_2). \tag{5.87}$$

It is called coercive if

$$\lim_{\|x\|\to\infty} h(x) = -\infty. \tag{5.88}$$

The following theorem is essential for proving existence of the solution in screening models:

Theorem 147 *Let Ω be a convex, open, bounded subset of R^m and, $F \subset H^1(\Omega)$ be a closed, convex set and $\phi : F \to R$ be a continuous, concave, coercive functional. Then the problem*

$$\max_{u\in F} \phi(u) \tag{5.89}$$

has a solution.

We again refer an interested reader to Kinderlehrer and Stampacchia (1980) for a proof of this result.

5.5 Optimization by Vector Space Methods

Let X and Z be a Banach spaces, $f : X \to R$ be a real valued functional and $g : X \to Z$ be a mapping. Moreover, assume Z is partially ordered by $\preceq$ and the partial order is consistent with the linear structure in the following sense

$$(z_1 \preceq z_2, z_3 \preceq z_4) \Rightarrow (z_1 + z_3 \preceq z_2 + z_4) \tag{5.90}$$

$$(z_1 \preceq z_2) \Rightarrow (\lambda z_1 \preceq \lambda z_2) \tag{5.91}$$

for $\forall z_1, z_2, z_3, z_4 \in Z$ and $\forall \lambda \geq 0$. I will call such Z an *ordered Banach space*. Let us consider the following optimization problem:

$$\max f(x) \tag{5.92}$$

$$s.t. g(x) \preceq \theta. \tag{5.93}$$

Assume that the maximum is achieved at $x_0 \in X$ and both $f(\cdot)$ and $g(\cdot)$ are continuously Frechet differentiable at x_0. If both X and Z are finite-dimensional, $\preceq$ is defined coordinate-wise, i. e. $\mathbf{a} \preceq \mathbf{b}$ if and only if

$$a_i \leq b_i \tag{5.94}$$

for all i. Assume that the Jacobi matrix $D\mathbf{g}/D\mathbf{x}$ evaluated at point $\mathbf{x}_0$ has full rank. Then there exists a vector $\boldsymbol{\lambda} \in R^m$ (here $n = \dim Z$) such that

$$\frac{\partial f}{\partial x_i}(\mathbf{x}_0) = \sum_{j=1}^{m} \lambda_j \frac{\partial g_j}{\partial x_i}(\mathbf{x}_0), \tag{5.95}$$

$$\boldsymbol{\lambda} \geq 0, \ \boldsymbol{\lambda} \cdot (\mathbf{g}(\mathbf{x}) - \boldsymbol{\theta}) = 0. \tag{5.96}$$

This statement is known as the Kunh-Tucker theorem. Intuitively, equations (5.95), (5.96) state that the gradient of the objective function should look in a direction in which all the constraints are increasing, since otherwise one can move in a direction that will leave the choice variable $\mathbf{x}$ within the constraint set, but increase the value of the objective.

It turns out that a similar statement is true in the general case. To formulate the theorem I first have to introduce a partial order on the dual of the ordered Banach space. Let Z be an ordered Banach space and $z^* \in Z^*$. I will define a partial order on Z^* in the following way

$$(z^* \overset{*}{\preceq} 0) \Leftrightarrow (\forall z \in Z, (z \preceq 0) \Rightarrow (z^*(z) \geq 0)) \tag{5.97}$$

$$(z_1^* \overset{*}{\preceq} z_2^*) \Leftrightarrow (z_1 - z_2 \overset{*}{\preceq} 0). \tag{5.98}$$

It is straightforward (and is left as an exercise for the reader) to check that $\overset{*}{\preceq}$ is consistent with the linear structure on Z^*. Now we are ready to formulate our main result.

Theorem 148 *Let X be a Banach space and Z be an ordered Banach space. Let $f : X \to R$ be a Gateaux differentiable functional and $g : X \to Z$ a Gateaux differentiable operator. Consider a problem*

$$\max f(x) \tag{5.99}$$

$$s.t. g(x) \preceq \theta. \tag{5.100}$$

Let $x_0 \in X$ solve (5.99)-(5.100) and assume that there exists $h \in X$ such that

$$g(x_0) + \frac{\partial g}{\partial h}(x_0)h < 0. \tag{5.101}$$

Then there exists $z^* \in Z$, $0 \overset{*}{\preceq} z^*$, *such that* x_0 *solves the unconstraint problem*

$$\max(f(x) + z^* g(x)) \tag{5.102}$$

and

$$z^*(g(x) - \theta) = 0. \tag{5.103}$$

I will not prove this theorem here. An interested reader is referred to Luenberger (1969). The following corollary is particularly important for the screening models.

Corollary 149 *Let* $\Omega \subset R^n$ *be a convex open bounded set. Consider the following problem*

$$\max_{u \in H^1(\Omega)} \int_\Omega f(\mathbf{x}, u, \mathbf{z}) \mathbf{dz} \tag{5.104}$$

$$s.t.\ Lu = \phi(\mathbf{x}, \mathbf{z}, u),\ u(\mathbf{z}) \geq u_0(\mathbf{z}), \tag{5.105}$$

where L *is some linear differential operator. Let* $u^*(\cdot)$ *be its solution. Then there exists a distribution* $\eta \geq 0$ *such that*

$$\eta(u - u_0) = 0 \tag{5.106}$$

and $u^*(\cdot)$ *solves*

$$\max_{u \in H^1(\Omega)} \int_\Omega (f(\mathbf{x}, u, \mathbf{z}) + \boldsymbol{\eta}(z) u(z)) \mathbf{dz} \tag{5.107}$$

$$s.t.\ Lu = \phi(\mathbf{x}, \mathbf{z}, u). \tag{5.108}$$

As we will see from the examples in the next part of the book, in general one cannot expect $\eta(\cdot)$ to be a function. One should also bear in mind that

$$\int_\Omega \boldsymbol{\eta}(\mathbf{z}) u(\mathbf{z}) \mathbf{dz} \tag{5.109}$$

is simply a convenient notation for $\eta(u)$and the product under the sign of the integral has a literal meaning only at the points at which the distribution is regular.

5.6 Calculus of Variation Problem with Convexity Constraints

In the previous section we discussed how to incorporate constraints on the state variable of a type

$$\mathrm{g}(u) \leq 0 \tag{5.110}$$

into the calculus of variation and the optimal control problems. We saw that a way for doing this is a generalization a well-known finite dimensional Kunh-Tucker conditions. In screening applications we will also meet constraints of a different nature.

Let $\Omega \subset R^m$ be an open, bounded, convex, set with a piece-wise smooth boundary. We will often face with a problem

$$\max \int_{\Omega} L(u, \nabla u, \mathbf{x}) d\mathbf{x} \tag{5.111}$$

$$s.t. \ u \text{ is } v - \text{convex}, \tag{5.112}$$

for some function $v(\alpha, \mathbf{x})$. In this Section I will explain how to put a Lagrange multiplier on the constraint (5.112), when v is linear in α, i.e. in the case when $v-$convexity is reduced to the ordinary convexity. The first order characterization of the solution is given by the following Theorem:

Theorem 150 *(Lions, 1998). Let v be linear in α and assume u^* solves problem (5.111)-(5.112). Then there exists a positively semidefinite matrix of distributions $\mu = \{\mu_{ij}\}$ such that*

$$\sum_{i=1}^{m} \frac{\partial}{\partial x_i}\left(\frac{\partial L}{\partial u_{x_i}}\right) = \frac{\partial L}{\partial u} - \boldsymbol{\lambda}\frac{\partial \mathbf{g}}{\partial u} + \sum_{i,j=1}^{m} \frac{\partial^2 \mu_{ij}}{\partial x_i \partial x_j}, \tag{5.113}$$

$$\int_{\Omega} \sum_{i,j=1}^{m} \mu_{ij} \frac{\partial^2 u_{ij}}{\partial x_i \partial x_j} d\mathbf{x} = 0, \ \sum_{i=1}^{m} \frac{\partial L}{\partial u_{x_i}} n_i(x) = 0 \ on \ \partial\Omega, \tag{5.114}$$

where $\mathbf{n}(\mathbf{x})$ is the unit vector normal to the boundary at $\mathbf{x}$.

5.7 Supermodularity and Monotone Comparative Statics

Consider the following maximization problem

$$\max_{\mathbf{x} \in X} u(\boldsymbol{\alpha}, \mathbf{x}). \tag{5.115}$$

First suppose that the solution is unique for every $\boldsymbol{\alpha}$. Under which conditions on set X and function $u(\cdot, \cdot)$ does the solution increases in $\boldsymbol{\alpha}$? If the set of maximizers is not a singleton for some values of $\boldsymbol{\alpha}$, under which conditions this set is still increasing in some sense with $\boldsymbol{\alpha}$ and in what sense does it increase?

It turns out that it is possible to provide an answer to these question provided that X is a lattice and $u(\cdot, \cdot)$ is supermodular in all its arguments. If this is the case the maximizing correspondence $\mathbf{x}(\boldsymbol{\alpha})$ is increasing in the strong set order, in particular if $\mathbf{x}(\boldsymbol{\alpha})$ is a function it is increasing. Below I am going to define and explain all the terms I introduced in this paragraph. The treatment here is necessarily brief. For a detailed treatment of the issue related to supermodularity and monotone comparative statics, see Topkis (1998).

Recall that a reflective and transitive relation $\succeq$ on set Ω is called a *partial order.* A set with a partial order on it is called a partially ordered set or a poset. Let $(\Omega, \succeq)$ be a poset. If $X \subset \Omega$ then $\xi \in \Omega$ is called an upper bound of X if $\xi \succeq x$ for any $x \in X$. An upper bound ξ of X is called the lowest upper bound or a *supremum* of X if $\xi \succeq \zeta$ for any other upper bound ζ. In a similar way one can define the biggest lower bound or the *infimum*. Now we are ready to give the following definition.

Definition 151 *A poset is called a lattice if every pair of elements has a supremum also known as their joint and an infimum also known as their meet.*

We will use the following notation:

$$\sup\{x, y\} = x \vee y, \;\; \inf\{x, y\} = x \wedge y. \tag{5.116}$$

Let us consider some examples of the lattices.

Example 152 *Let $\Omega = R^n$ and $\succeq$ is a coordinate-wise partial order, i. e. $(\mathbf{x} \succeq \mathbf{y}) \Leftrightarrow (x_i \geq y_i$ for all $i)$. Then $(\Omega, \succeq)$ is a lattice and*

$$x \vee y = (\max\{x_1, y_1\}, ..., \max\{x_n, y_n\}) \tag{5.117}$$

$$x \wedge y = (\min\{x_1, y_1\}, ..., \min\{x_n, y_n\}). \tag{5.118}$$

Example 153 *Let $\Omega = C([0,1])$ and $f \succeq g$ if and only if $f(x) \geq g(x)$ for any $x \in [0,1]$. Then $(\Omega, \succeq)$ is a lattice and*

$$f \vee g = h_1, \;\; x \wedge y = h_2, \tag{5.119}$$

where continuous functions h_1 and h_2 are defined by

$$h_1(x) = \max(f(x), g(x)) \tag{5.120}$$

$$h_2(x) = \min(f(x), g(x)). \tag{5.121}$$

Example 154 *Let X be any set and $\Omega = 2^X$ is the set of its subsets. For any $A, B \in \Omega$ (namely, $A \subset X$ and $B \subset X$) define a partial order $\succeq$ by $A \succeq B$ if and only if $A \subset B$. Then $(\Omega, \succeq)$ is a lattice and*

$$A \vee B = A \cup B \tag{5.122}$$

$$A \wedge B = A \cap B. \tag{5.123}$$

Any lattice $(\Omega, \succeq)$ induces a relation on its subsets known as the *strong set order.*

Definition 155 *Let $(\Omega, \succeq)$ be a lattice and $X_1, X_2 \in 2^\Omega$. Then we will say that X_1 dominates X_2 in strict set order and write*

$$X_1 \succeq_S X_2 \tag{5.124}$$

if for any $x_1 \in X_1$ and any $x_2 \in X_2$

$$x_1 \vee x_2 \in X_1, \ x_1 \wedge x_2 \in X_2. \tag{5.125}$$

This definition implies that if x_1 and x_2 are ordered by $\succeq$ then X_1 always contains the largest element, while X_2 contains the smallest. For example, if Ω is the real line ordered in the usual way and $X_1 \succeq_S X_2$ then the set X_2/X_1 (the set of elements of X_2, which do not belong to X_1) will lie to the left of $X_1 \cap X_2$, which lies to the left of the set X_1/X_2.

Let Ω be a lattice and $f : \Omega \to R$ be a real valued function.

Definition 156 *Function f is called supermodular if $\forall x_1, x_2 \in \Omega$*

$$f(x_1 \vee x_2) + f(x_1 \wedge x_2) \geq f(x_1) + f(x_2). \tag{5.126}$$

The next example allows us to reveal the economic meaning of supermodularity.

Example 157 *Let $f : R^2 \to R$ be a twice differentiable function and let $\succeq$ be the partial coordinate-wise order on R^2. First, assume that $f(\cdot)$ is supermodular and consider points $x_1 = (x, y)$ and $x_2 = (x^*, y+\varepsilon)$ for some $\varepsilon > 0$ and some $x^* < x$. Then (5.126) implies that*

$$f(x, y+\varepsilon) - f(x, y) \geq f(x^*, y+\varepsilon) - f(x^*, y). \tag{5.127}$$

Dividing by ε and taking ε to zero one obtains

$$f_y(x, y) \geq f_y(x^*, y) \tag{5.128}$$

for all $x^* < x$. That is the partial derivative f_y is increasing in x, therefore $f_{xy} \geq 0$.

Now let us assume that $f_{xy} \geq 0$. Let $x_1 = (x, y)$, $x_2 = (x^*, y^*)$ and without loss of generality assume that $x \geq x^*$. Then for any ξ, function $f_y(\cdot, \xi)$ is increasing and therefore

$$f_y(x, \xi) \geq f_y(x^*, \xi). \tag{5.129}$$

But then

$$\int_y^{\max\{y,y^*\}} f_y(x, \xi) d\xi \geq \int_{\min\{y,y^*\}}^{y^*} f_y(x^*, \xi) d\xi. \tag{5.130}$$

Indeed, either $\max\{y, y^*\} = y$ *in which case* $\min\{y, y^*\} = y^*$ *and integrals on both sides vanish, or* $\max\{y, y^*\} = y^*$ *in which case* $\min\{y, y^*\} = y$ *and* *the inequality follows from integration of (5.129) from y to y^*, where $y^* \geq y$. Performing integration in (5.130) and rearranging terms one obtains*

$$f(x, \max\{y, y^*\}) + f(x^*, \min\{y, y^*\}) \geq f(x, y) + f(x^*, y^*), \qquad (5.131)$$

therefore $f(\cdot)$ is supermodular. We conclude that $f(\cdot)$ is supermodular if and only if

$$f_{xy} \geq 0. \qquad (5.132)$$

Note also that we did not use differentiability in x of $f_y(\cdot, y)$ in any essential way. Hence, we can reformulate the result: $f(\cdot)$ is supermodular if and only if f_y increases in x (or f_x increases in y).

Suppose $f(\cdot)$ is a production function. Then equation (5.132) implies that the marginal product of input x increases in y and vice versa, therefore inputs x and y are complimentary. Therefore, from an economic point of view supermodularity reflects the idea of complementarity.

The importance of the idea of supermodularity for the screening applications comes from the following monotone maximum theorem.

Theorem 158 *Assume that X and Ω are lattices and $u : X \times \Omega \to R$ is a supermodular function. For every $\alpha \in \Omega$ consider a problem*

$$\max_{x \in X} u(\alpha, x). \qquad (5.133)$$

Assume that for any $\alpha \in \Omega$ a solution to problem (5.133) exists, that is the correspondence

$$\phi(\alpha) = \arg\max_{x \in X} u(\alpha, x) \qquad (5.134)$$

is not empty-valued for all $\alpha \in \Omega$. Then hat $\alpha_2 \succeq \alpha_1$ implies that $\phi(\alpha_2) \succeq_S \phi(\alpha_1)$.

Proof. Let $\alpha_1, \alpha_2 \in \Omega$ and $\alpha_2 \succeq \alpha_1$. Let $x_1 \in \phi(\alpha_1)$ and $x_2 \in \phi(\alpha_2)$. Since X is a lattice, $x_1 \wedge x_2 \in X$ and $x_1 \vee x_2 \in X$. Let us prove that $x_1 \vee x_2 \in \phi(\alpha_2)$ and $x_1 \wedge x_2 \in \phi(\alpha_1)$. For this purpose, consider the following inequalities, implied by the supermodularity of $u(\cdot)$ and the observation that $\alpha_1 \vee \alpha_2 = \alpha_2$ and $\alpha_1 \wedge \alpha_2 = \alpha_1$:

$$\begin{aligned} u(x_1 \vee x_2, \alpha_2) - u(x_2, \alpha_2) &\geq u(x_1, \alpha_2) - u(x_1 \wedge x_2, \alpha_2) \qquad (5.135) \\ u(x_1, \alpha_2) - u(x_1 \wedge x_2, \alpha_2) &\geq u(x_1, \alpha_1) - u(x_1 \wedge x_2, \alpha_1). \qquad (5.136) \end{aligned}$$

Therefore,

$$u(x_1 \vee x_2, \alpha_2) - u(x_2, \alpha_2) \geq u(x_1, \alpha_1) - u(x_1 \wedge x_2, \alpha_1) \qquad (5.137)$$

But since x_i is the maximizer for the respective value α_i

$$\begin{cases} u(x_1 \vee x_2, \alpha_2) - u(x_2, \alpha_2) \leq 0 \\ u(x_1, \alpha_1) - u(x_1 \wedge x_2, \alpha_1) \geq 0 \end{cases}. \qquad (5.138)$$

Combining (5.138) and (5.137) one obtains

$$0 \geq u(x_1 \vee x_2, \alpha_2) - u(x_2, \alpha_2) \geq u(x_1, \alpha_1) - u(x_1 \wedge x_2, \alpha_1) \geq 0. \quad (5.139)$$

Therefore,

$$u(x_1 \vee x_2, \alpha_2) = u(x_2, \alpha_2) \quad (5.140)$$

$$u(x_1, \alpha_1) = u(x_1 \wedge x_2, \alpha_1). \quad (5.141)$$

Hence, $x_1 \vee x_2 \in \phi(\alpha_2)$ and $x_1 \wedge x_2 \in \phi(\alpha_1)$. ■

One can prove that in the conditions of the above theorem the maximizing correspondence allows for an increasing selection. The proof of this result is left as an exercise to the reader. Note that assuming that the solution to the maximization problem exists the monotone maximum theorem did not make any assumptions about continuity of differentiability of the objective function (of course, such assumptions might be necessary to ensure the existence of the solution). Therefore, it can be applied also in the case when either the space of maximizers or the parameters space is discrete.

Example 159 *In a first-price private value auction, each bidder has a value α for the object auctioned, which is her private information. The highest bidder wins and pays her bid. If she bids amount x and is successful she receives utility $u(\alpha - x)$, where $u(\cdot)$ is some twice-differentiable strictly concave function. Let $p(x)$ be the probability of winning, when the bid is x. Then the ex-ante expected utility of the bidder is*

$$v(\alpha, x) = p(x)u(\alpha - x) \quad (5.142)$$

Since the probability of winning is increasing function of x, function $v_\alpha(\alpha, \cdot)$ defined by

$$v_\alpha = p(x)u'(\alpha - x) \quad (5.143)$$

is an increasing function of x. Therefore, the bidder's objective is supermodular and her bid is increasing in x. Note that this conclusion is independent of the distribution of values among the bidders.

5.8 Hausdorff Metric on Compact Sets of a Metric Space

Let (X, d) be a metric space. For any $A \subset X$ and any $x \in X$ define distance from point x to set A by

$$d(x, A) = \inf_{y \in A} d(x, y). \quad (5.144)$$

Proposition 160 *Let $A \subset X$ be a closed set. Then $d(x, A) = 0$ if and only if $x \in A$.*

Proof. It is clear that $x \in A$ implies $d(x, A) = 0$. No suppose $x \notin A$. Since A is closed its complement is open, therefore $\exists \varepsilon > 0$ such that

$$\{y \in X : d(x, y) < \varepsilon\} \cap X = \varnothing. \tag{5.145}$$

Therefore,

$$d(x, A) \geq \varepsilon > 0. \tag{5.146}$$

■

Definition 161 *Set $K \subset X$ is called compact if for any family of open set U_α, $\alpha \in F$ such that*

$$K \subset \bigcup_{\alpha \in F} U_\alpha \tag{5.147}$$

there exist a finite set $\{\alpha_1, ..., \alpha_n\} \subset F$ such that

$$K \subset \bigcup_{i=1}^{n} U_{\alpha_i}. \tag{5.148}$$

In words, a set is compact if any its open cover has a finite subcover. An important fact about compact sets is given by the following proposition.

Proposition 162 *Every compact subset of a metric space is closed.*

This fact is standard and its proof can be found in any book on general topology. Therefore, I omit the proof here. Let $H(X)$ be the set of compact non-empty subsets of X. For any $A \in H(X)$ and any $r > 0$ define $N_r(A)$, the open $r-$neighborhood of A, as follows:

$$N_r(A) = \{y \in X : \exists x \in A : d(x, y) < r\}. \tag{5.149}$$

For any $A, B \in H(X)$ the Hausdorff metric is defined in the following way:

$$h(A, B) = \inf_{r>0} \{r : A \subset N_r(B) \& B \subset N_r(A)\}. \tag{5.150}$$

Let us also define $\inf \varnothing = \infty$.

Proposition 163 *Equation (5.150) defines a metric on $H(X)$. This metric is known as Hausdorff metric.*

Proof. It is clear from the definition that $h(A, B) \geq 0$ and $h(A, B) = h(B, A)$. Since $A \subset N_r(A)$ for all $r > 0$ we also have $h(A, A) = 0$. Assume that $h(A, B) = 0$ for some $A, B \in H(X)$. We have to prove that $A = B$. Let us first prove that $A \subset B$. For this purpose take $\forall x \in A$. Note that $h(A, B) = 0$ implies that since $A \subset N_r(B)$ for $\forall r > 0$ and therefore $x \in$

$N_r(B)$. Hence, $d(x, B) < r$ for any $r > 0$ which implies $d(r, B) = 0$. Since X is compact and therefore closed $x \in B$, therefore $A \subset B$. Similarly, $B \subset A$ and hence, $A = B$. Finally, we have to check the triangle inequality. Let $A, B, C \in H(X)$ and take any $r > 0$. For $\forall x \in A$ there exists $y \in B$ such that $d(x, y) < h(A, B) + r$ and there exists $z \in C$ such that $d(y, z) < h(B, C) + r$. Therefore,

$$A \subset N_{h(A,B)+h(B,C)+2r}(C). \tag{5.151}$$

Similar reasoning proves that

$$C \subset N_{h(A,B)+h(B,C)+2r}(A). \tag{5.152}$$

Since this is true for any $r > 0$ one obtains

$$h(A, C) \leq h(A, B) + h(B, C). \tag{5.153}$$

■

Therefore, we proved that ($H(X)$, h) is a metric space. It can also be shown that if the underlying space (X, d) is complete so is ($H(X)$, h). An important fact about this space is given by the following lemma.

Lemma 164 *Let* $\{A_n\}_{n=0}^{\infty}$ *be a family of compact sets of* X *and let* $A_{n+1} \subset A_n$ *for all* $n \in N$. *Then*

$$\lim_{n \to \infty} A_n = A = \bigcap_{n=0}^{\infty} A_n, \tag{5.154}$$

where the limit is understood as the limit in ($H(X)$, h)).

Proof. Take $\forall \varepsilon > 0$. We have to prove that $\exists n_0 \in N$ such that $h(A, A_n) < \varepsilon$ for $\forall n > n_0$. Since $A \subset A_n$, we conclude that $A \subset N_\varepsilon(A_n)$. To prove that $A_n \subset N_\varepsilon(A)$ for a sufficiently big n first note that $N_\varepsilon(A)$ is an open set and so is $X \backslash A_n$ for any n, since A_n is closed. Therefore the family

$$\{N_\varepsilon(A)\} \cup \{X \backslash A_n, n \in N\} \tag{5.155}$$

is an open cover of A_1. Since A_1 is compact, there exists $n_0 \in N$ such that

$$A_1 \subset N_\varepsilon(A) \bigcup_{n=0}^{n_0} (X \backslash A_n) = N_\varepsilon(A) \cup (X \backslash A_{n_0}). \tag{5.156}$$

Therefore, for $\forall n > n_0$

$$A_1 \subset N_\varepsilon(A) \cup (X \backslash A_n). \tag{5.157}$$

Now assume $\exists x \in A_n$ such that $x \notin N_\varepsilon(A)$. But then $x \in A_1$ (since $A_n \subset A_1$) but $x \notin N_\varepsilon(A) \cup (X \backslash A_n)$, which contradicts (5.157). Therefore, $A_n \subset N_\varepsilon(A)$ and $h(A, A_n) < \varepsilon$. ■

Now let $X = R^n$ and consider a sequence of sets $\{A_n\}$ in $H(R^n)$ converging to a singleton set $\{\mathbf{a}\}$ for some $\mathbf{a} \in R^n$. Our next objective is to prove that their volumes converge to zero, where volume of a compact set is by definition its Lebesgue measure.

Lemma 165 *Let $A_n \in H(R^n)$ and sequence $\{A_n\}_{n=0}^{\infty}$ converges to $\{\mathbf{a}\}$ in $(H(R^n), h)$ for some $\mathbf{a} \in R^n$. Define*

$$V(A_n) = \int_{A_n} 1 \cdot \mathbf{dx}. \tag{5.158}$$

Then

$$\lim_{n \to \infty} V(A_n) = 0. \tag{5.159}$$

Proof. First note that for any $\varepsilon > 0$ there exists $\delta > 0$ such that

$$V(B(\mathbf{a},\delta)) < \varepsilon. \tag{5.160}$$

Since$\{A_n\}_{n=0}^{\infty}$ converges to $\{\mathbf{a}\}$ in $(H(R^n), h)$, there exists $n_0 \in N$ such that $A_n \subset B(\mathbf{a},\delta)$ for $\forall n > n_0$. Therefore,

$$V(A_n) \leq V(B(\mathbf{a},\delta)) < \varepsilon, \tag{5.161}$$

i. e.

$$\lim_{n \to \infty} V(A_n) = 0. \tag{5.162}$$

■

Now let $C \in H(R^n)$ be a convex set. Define its surface area, $S(C)$ as the following limit

$$S(C) = \lim_{\varepsilon \to 0} \frac{V(C + \varepsilon B) - V(C)}{\varepsilon}, \tag{5.163}$$

here

$$C + \varepsilon B = \{\mathbf{x} \in R^n : \exists \mathbf{u} \in C \text{ and } \mathbf{v} \in B : \mathbf{x} = \mathbf{u} + \varepsilon \mathbf{v}\} \tag{5.164}$$

$$B = \{\mathbf{x} \in R^n : \sum_{i=1}^{n} x_i^2 \leq 1\}. \tag{5.165}$$

It can be shown that limit (5.163) exists for any convex compact C.

Definition 166 *A set $F \subset R^n$ is said to have full dimension in R^n if for any $\mathbf{x} \in F$ and $\forall \varepsilon > 0$ there exist $\mathbf{x}_1, ..., \mathbf{x}_n \in F \cap B(\mathbf{x}, \varepsilon)$ such that vectors $\mathbf{x}_1 - \mathbf{x}, ..., \mathbf{x}_n - \mathbf{x}$ are linearly independent.*

The following result states that if a sequence of sets converges to a point in Hausdorff metric then their surface area converges to zero, provided $n > 1$.

Lemma 167 *Assume that $A_n \in H(R^n)$ for some $n > 1$, A_n is convex for all n, has full dimension in R^n and sequence $\{A_n\}_{n=0}^{\infty}$ converges to $\{\mathbf{a}\}$ in $(H(R^n), h)$ for some $\mathbf{a} \in R^n$. Then*

$$\lim_{n \to \infty} S(A_n) = 0. \tag{5.166}$$

Proof. Take any $\varepsilon > 0$. Since$\{A_n\}_{n=0}^{\infty}$ converges to $\{\mathbf{a}\}$ in $(H(R^n), h)$, there exists $n_0 \in N$ such that $A_n \subset B(\mathbf{a},\varepsilon)$ and $A_n + \varepsilon B \subset B(\mathbf{a},2\varepsilon)$ for $\forall n > n_0$. For any $B \in H(R^n)$ define

$$d(B) = \sup_{(\mathbf{x},\mathbf{y}) \in B \times B} d(x, y). \tag{5.167}$$

It is easy to see that for any convex $B \in H(R^n)$ of full dimension

$$V(B) = O([d(B)]^n). \tag{5.168}$$

Since $d(A_n) \leq \varepsilon$ and $d(A_n + \varepsilon B) \leq 2\varepsilon$ and $V(A_n + \varepsilon B) > V(A_n)$ one obtains

$$0 < \frac{V(C + \varepsilon B) - V(C)}{\varepsilon} \leq O(\varepsilon^{n-1}), \tag{5.169}$$

therefore

$$\lim_{\varepsilon \to 0} \frac{V(C + \varepsilon B) - V(C)}{\varepsilon} = 0 \tag{5.170}$$

for $n > 1$. ■

Finally, it is important to give the following definition for the future use:

Definition 168 *A set Ω is called strictly convex if for any $\alpha, \beta \in \Omega$ and any $\lambda \in (0, 1)$*

$$\gamma = \lambda\alpha + (1 - \lambda)\beta \in Int(\Omega), \tag{5.171}$$

i. e. there exists neighborhood $U(\gamma)$ of γ such that $U(\gamma) \subset \Omega$.

Intuitively, a set is strictly convex if its boundary does not contain parts of hyperplanes.

5.9 Generalized Envelope Theorems

An envelope theorem is a statement that links the derivative of the value function with respect to the parameters of the model to the derivative of the objective evaluated at the optimal choice. The traditional formulation is the following:

Theorem 169 *Let $Z \subset R^m$ be an open non-empty set and define the value function $V(\cdot) : Z \to R$ by*

$$V(\mathbf{z}) = \max_{\mathbf{x} \in R^n} u(\mathbf{x}, \mathbf{z}) \tag{5.172}$$

Assume that both u is a differentiable function of its arguments and the maximum is achieved for some $\mathbf{x} = \mathbf{x}(\mathbf{z})$. Then $V(\cdot)$ is differentiable and

$$\frac{\partial V}{\partial z_i}(\mathbf{z}) = \frac{\partial u}{\partial z_i}(\mathbf{x}(\mathbf{z}), \mathbf{z}), \tag{5.173}$$

Proof. I skip the proof of differentiability of $V(\cdot)$. To establish (5.173), note that

$$V(\mathbf{z}) = u(\mathbf{x}(\mathbf{z}), \mathbf{z}) \tag{5.174}$$

and using the chain rule

$$\frac{\partial V}{\partial z_i}(\mathbf{z}) = \frac{\partial u}{\partial z_i}(\mathbf{x}(\mathbf{z}), \mathbf{z}) + \sum_{j=1}^{n} \frac{\partial u}{\partial x_j}(\mathbf{x}(\mathbf{z}), \mathbf{z}) \frac{\partial x_j}{\partial z_i}. \tag{5.175}$$

The first order conditions for the constraint optimization imply that

$$\frac{\partial u}{\partial x_j}(\mathbf{x}(\mathbf{z}), \mathbf{z}) = 0, \tag{5.176}$$

therefore

$$\frac{\partial V}{\partial z_i}(\mathbf{z}) = \frac{\partial u}{\partial z_i}(\mathbf{x}(\mathbf{z}), \mathbf{z}). \tag{5.177}$$

and (5.173) is established. ■

Note that though the proof relied on the differentiability of the objective and the constraint with respect to $\mathbf{x}$, for formula (5.173) to make sense $V(\cdot)$ and $u(\cdot, \mathbf{x})$ should only be differentiable with respect to $\mathbf{z}$. Milgrom and Segal (2002) showed that the assumption of differentiability with respect to $\mathbf{x}$ is indeed dispensable, more precisely the following theorem holds.

Theorem 170 *Let $Z \subset R^m$ be an open non-empty set and X be an arbitrary set. Define the value function $V(\cdot) : Z \to R$ by*

$$V(\mathbf{z}) = \max_{\mathbf{x} \in X} u(\mathbf{x}, \mathbf{z}) \tag{5.178}$$

Assume that $u(\cdot, \mathbf{x})$ is differentiable for $\forall \mathbf{x} \in X$ and the maximum is achieved for some $\mathbf{x} = \mathbf{x}(\mathbf{z})$. Then $V(\cdot)$ is almost everywhere differentiable and at the differentiability points

$$\frac{\partial V}{\partial z_i}(\mathbf{z}) = \frac{\partial u}{\partial z_i}(\mathbf{x}(\mathbf{z}), \mathbf{z}). \tag{5.179}$$

We will not prove this theorem here, the proof relies on the idea of absolute continuity not discussed in this book. An interested reader is referred to Milgrom and Segal (2002).

5.10 Exercises

1. Prove that functional $\varphi_{s,\mathbf{a}} \in D$, where $\varphi_{s,\mathbf{a}}$ is defined by (5.20).
2. Prove that if $\varphi_{s,\mathbf{a}}$ is defined by (5.20) then there exists no Lebesgue integrable function $f : R^n \to R$ such that

$$\varphi_{s,\mathbf{a}}(h) = \int_{R^n} f(\mathbf{x})h(\mathbf{x})\mathbf{dx}. \tag{5.180}$$

3. Let $u : R^n \to R$ be twice differentiable. Consider a coordinate-wise ordering on R^n. Prove that $u(\mathbf{x})$ is supermodular if and only if

$$\frac{\partial^2 u}{\partial x_i \partial x_j}(\mathbf{x}) \geq 0 \tag{5.181}$$

for all $\mathbf{x} \in R^n$ and all $i \neq j$.
4. Prove that if Ω and X are lattices and $\phi : X \to \Omega$ is a correspondence such that $x_2 \succeq x_1$ implies that $\phi(x_2) \succeq_S \phi(x_1)$ then there exists function $h(x)$ such that $h(x) \in \phi(x)$ for any x and $x_2 \succeq x_1$ implies that $h(x_2) \succeq h(x_1)$, i. e. correspondence $\phi(\cdot)$ allows an increasing selection.
5. Prove that for any $r > 0$ set $N_r(A)$ defined by (5.149) is open.
6. Prove that (5.155) indeed defines a covering of A_1. Hint: use de-Morgans formula

$$\bigcup_{n=1}^{\infty}(X \backslash A_n) = X \backslash \bigcap_{n=1}^{\infty} A_n. \tag{5.182}$$

7. Prove that if $C \subset R^n$ is a convex compact set then limit (5.163) exists.
8. Let $B \subset R^n$ be a convex compact set of full dimension. Prove formula (5.168).
9. Let $B \subset R^n$ be a strictly convex compact set and $u(\cdot)$ a strictly increasing, continuous, convex function. Moreover, assume that

$$\min_{\mathbf{x} \in B} u(\mathbf{x}) = 0. \tag{5.183}$$

For any $\varepsilon > 0$ define

$$B(\varepsilon) = \{\mathbf{x} \in B : u(\mathbf{x}) \leq \varepsilon\}. \tag{5.184}$$

Prove that:

a). Set $B(0)$ is a singleton, i. e. $\exists \mathbf{b} \in B$ such that $B(0) = \{\mathbf{b}\}$;

b). if $\{\varepsilon_n\}$ converges to ε that $B(\varepsilon_n) \to B(\varepsilon)$ in $(H(R^n), h)$, i. e. mapping from the real line into the space convex compact sets endowed with Hausdorff metric given by (5.184) is continuous;

c). Let $n > 1$, then

$$\lim_{\varepsilon \to 0} V(B(\varepsilon)) = \lim_{\varepsilon \to 0} S(B(\varepsilon)) = 0. \tag{5.185}$$

5.11 Bibliographic Notes

Obviously, there is no single source for the material covered in this chapter. For the first section an excellent and an up-to-date reference is Vladimirov (2002). The reader should be warned that in that book the distributions are called *generalized functions,* which is the verbatim translation of the Russian term for a distribution. Kinderlehrer and Stampacchia (1980) is the classical text on the obstacle problems in mathematical physics and is the leading reference for Sections 2 and 3. Luenberger (1969) contains a very detailed exposition of the material of Section 4 with good examples of applications of these techniques to the optimal control problems. The difference of the exposition in Luenberger (1969) from the one given here is that he uses the measure-theoretic rather than the distribution-theoretic language.

The material of the first four sections is rather standard and can be found at many other books apart from the ones cited above. An exposition of the material of Section 5 is harder to find in the literature. The only source I am aware of is Meyer (1966). Note that the later edition of this book (Dellacherie and Meyer, 1978) does not contain this material and the 1966 edition became a bibliographic rarity. Section 6 reviews a recent paper published by Lions (1988). The paper is in French with an English annotation and no other source of these results is available.

A good exposition for the economists of the material covered in Section 7 is Topkis (1998). A more formal and concise exposition can be found in Carter (2001). A classical but still very useful book of Eggleston (1958) is the best reference for Section 8.

Material in Section 9 is based on recent journal papers by Milgrom and Segal (2002) and Sah and Zhao (1998). One does not need any special preparation to understand the formulation of the theorems and the intuition behind them. To understand the proofs, however, the reader might want to consult a text in real analysis. A useful book for this purpose will be Royden (1988).

Part II

Economics of Multidimensional Screening

In this part I am going to review known results and present some new developments in the theory of multidimensional screening. I start with the well-known unidimensional model, considered first by Mussa and Rosen (1978). Though the material in that section is rather standard, I make a special emphasis on comparison of the relative power of three approaches: *direct*, *dual*, and *Hamiltonian*. I emphasize that even in the unidimensional case the Hamiltonian approach is the most powerful and allows us to arrive at the solution of the complete problem without any additional assumptions. Therefore, I use it to derive all known results: the solution to the relaxed problem, Mussa and Rosen's ironing conditions, solution to the problem with type dependent (Jullien, 2000) and random (Rochet and Stole, 2001) participation constraints in a unified way.

The unidimensional models, however, do not cover all the situations of practical interest. Indeed, often the nonlinear tariffs specify payment as a function of a variety of characteristics. For example, railroad tariffs specify charges based on weight, volume, and distance of each shipment. Different customers may value each of these characteristics differently, hence the customer's type will not in general be captured by a unidimensional characteristic and a problem of multidimensional screening arises. In such models the consumer's private information (her type) is captured by an $m-$dimensional vector, while the good produced by the monopolist has n quality dimensions.

There is no a priori connection between m and n. Consider, for example, the case when the monopolist is a long distance company which allows its customers to make calls to two locations A and B. Assume that customers who have relatives at location A usually also have relatives at location B. Let τ_i be time spend by a customer on talks with location i and t is the amount paid by the customer to the company. Assume that her utility is $\alpha u(\tau_A, \tau_B) - t$, where function u is the same among the customers, and α is a privately observed type. In this case $n = 2$, while $m = 1$; hence $n > m$. For an example where $n < m$, consider a situation when a monopolist is a computer company. Suppose it designs computers and software. Consumers may differ in their preferences over a computer, a piece of software, and a bundle consisting of a computer and a piece of software. (For instance, their utility functions may be of the form $u(x_1, x_2) = \alpha_1 x_1 + \alpha_2 x_2 + \alpha_3 v(x_1, x_2)$, where x_1, x_2 are qualities of a computer and software respectively). In this case $n = 2$, while $m = 3$; hence $n < m$.

The multidimensional problems exhibit some new qualitative features not present in the unidimensional case. For example, in general there will be distortions of quality at the external boundary of the types set (Rochet and Chone, 1998), though under some technical assumptions there will still exist a type that is served efficiently (Basov, 2002).

On the technical side, it proved excessively cumbersome to generalize the direct approach to the multidimensional problems. The main obstacle is the presence of integrability constraints. These constraints ensure that it is possible to define the concept of the information rent similar to the unidimensional case. This motivated dual Rochet and Chone (1998) to develop the dual approach to multidimensional screening problems. They assumed that $m = n$ and preferences are linear in type. Basov (2001) generalized the dual approach to the case $n \geq m$ retaining the linearity assumption. It, however, becomes inapplicable if $m > n$ and very cumbersome for the nonlinear case even if $m = n$. The Hamiltonian approach, on the other hand, remains powerful even under rather general assumptions. It can also easily accommodate any exogenous constraints that may arise in different applications and some steps where taking en-route of using it to solve the complete problem in the general case.

This part is organized in the following way. Chapters 6 and 7 consider the unidimensional and multidimensional cases respectively, assuming that the consumer's preferences are quasilinear in money. This is the case that received most attention in the literature. In fact, I am not aware of any results in the multidimensional case that dispense with quasilinearity assumption. However, in a recent note on economics of the government grants distribution to the research institutions, Bardsley and Basov (2004) concluded that the most natural way to model this situation leads to a screening problem with two-dimensional type and one screening instrument where the preferences of the institutions are not quasilinear in money. Another example of a problem, which is not quasilinear in money is the optimal insurance problem, in which both the probability of loss and the degree of risk-aversion are private information of the consumer. The competitive version of this model, in which both parameters can take only two different values, high and low, was considered by Smart (2000) and Villeneuve (2003). The existence of situations of economic interest where the most natural model is not quasilinear motivates a general investigation of such models undertaken in Chapter 8. Chapter 9 collects all known existence, uniqueness, and regularity results. Chapter 10 concludes and outlines some directions for future development.

In this part I will refer to the results of the previous part in the following way: (I, Theorem 155) will mean Theorem 155 of part one, (I, 2.35) will mean formula 2.35 of part one, etc.

6
The Unidimensional Screening Model

In this chapter I concentrate on the case $m = n = 1$. As it will be clear from the exposition the crucial assumption is that $m = 1$ (the consumer's type is unidimensional). All the results will, however, easily go through if $n > 1$ (this is assumed in some papers. See, for example, Mirman and Sibley (1980), and Sappington, 1983)). I assume $n = 1$ only for the expositional simplicity.

Assume a monopolist who faces a continuum of consumers produces a good of quality x. The cost of production is assumed to be given by a strictly increasing, convex, twice differentiable function, $c(\cdot)$. Each consumer is interested in consuming at most one unit of the good and has a utility

$$u(\alpha, x) - t, \tag{6.1}$$

where α is her unobservable type distributed on $(0, 1)$ according to a strictly positive, continuous density function $f(\cdot)$, t is the amount of money transferred to the monopolist, and $u(\alpha, x)$ is a continuous function, strictly increasing in both arguments. Consumers have an outside option $u_0(\alpha)$. The monopolist is looking for a mechanism that would maximize her profits.

According to the *Revelation Principle* (see, for example, Mas-Colell, Whinston, and Green, 1995) one can without loss of generality assume that the monopolist uses direct revelation mechanisms, that is she announces quantity and price schedules $x(\alpha)$ and $t(\alpha)$ and allows the consumers to announce their type. Moreover, since incentive compatibility implies that the consumers that consume that same bundle should pay he same amount, one can without loss of generality assume that the monopolist simply announces a non-linear tariff $t(x)$. The last statement, known as the *taxation principle* was first formulated by Rochet (1985).

The above consideration can be summarized by the following model. The monopolist select a continuos function $t : R \to R$ to solve

$$\max_{t(\cdot)} \int_0^1 (t(x(\alpha)) - c(x(\alpha)))f(\alpha)d\alpha, \tag{6.2}$$

where $c(x)$ is the cost of producing a good with quality x and $x(\alpha)$ satisfies

$$x(\alpha) \in \arg\max(u(\alpha, x) - t(x)) \tag{6.3}$$
$$\max(u(\alpha, x) - t(x)) \geq 0. \tag{6.4}$$

Here I assumed that the utility of the outside option is type independent and normalized it to be zero. This model was first introduced by Mussa and Rosen (1978). We will start we a detailed study of its properties.

6.1 Spence-Mirrlees Condition and Implementability

Suppose you are given an allocation $x(\alpha)$. Under what conditions does there exist a non-linear tariff such that (6.3) is satisfied? If such a tariff exists the allocation is called implementable. Formally, the following definition holds.

Definition 171 *An allocation $x(\cdot)$ is called implementable if it is measurable and there exists a measurable function $t(\cdot)$ such that*

$$x(\alpha) \in \arg\max_{x \in R_+} (u(\alpha, x) - t(x)). \tag{6.5}$$

Our first objective is to characterized the set of implementable allocations. We will do this under an additional condition: Function $u(\alpha, x)$ is twice continuously differentiable and

$$u_{\alpha x} > 0. \tag{6.6}$$

Equation (6.6) is known as the Spence-Mirrlees condition of the single-crossing property. The last name reflects the fact that under (6.6) the indifference curves of the consumers of different types in (x, t) space cross only once. A well-known result is given by the following Theorem (Mussa and Rosen, 1978).

Theorem 172 *Allocation $x(\cdot)$ is implementable if and only if it is increasing.*

Note that if $u(\cdot,\cdot)$ satisfies the single crossing property then function $U(\cdot,\cdot)$ defined by

$$U(\alpha, x) = u(\alpha, x) - t(x) \tag{6.7}$$

it is supermodular in its arguments for any $t(\cdot)$. Therefore, the necessity of this result follows from the Monotone Maximum Theorem. (I, Theorem 155). Let us prove the sufficiency.

Proof. Suppose $x(\cdot)$ is increasing. Then it is measurable and we can define the tariff $t(\cdot)$ by

$$t(\beta) = u(\beta, x(\beta)) - \int_0^\beta u_\alpha(\gamma, x(\gamma))d\gamma. \tag{6.8}$$

Note that the pair of functions $x(\cdot)$ and $t(\cdot)$ define a nonlinear tariff $t(x)$ in a parametric form. Consider the decision of a consumer of type α. From her perspective, choosing quality x is equivalent to choosing type β she pretends to be (this statement is a particular case of the Revelation Principle, see Mas-Collel, Whinston, and Green (1995) for a discussion), therefore she solves

$$\max_{\beta\in[0,1]} (u(\alpha, x(\beta)) - u(\beta, x(\beta)) + \int_0^\beta u_\alpha(\gamma, x(\gamma))d\gamma). \tag{6.9}$$

Let

$$V(\alpha, \beta) = u(\alpha, x(\beta)) - u(\beta, x(\beta)) + \int_0^\beta u_\alpha(\gamma, x(\gamma))d\gamma. \tag{6.10}$$

Our objective is to prove that

$$V(\alpha, \alpha) \geq V(\alpha, \beta), \tag{6.11}$$

which implies that

$$\alpha \in \arg\max_{\beta\in[0,1]} (u(\alpha, x(\beta)) - u(\beta, x(\beta)) + \int_0^\beta u_\alpha(\gamma, x(\gamma))d\gamma), \tag{6.12}$$

and therefore tariff (6.8) implements allocation $x(\cdot)$.

To prove (6.11) note that

$$V(\alpha, \alpha) = \int_0^\alpha u_\alpha(\gamma, x(\gamma))d\gamma \tag{6.13}$$

and one can transform (6.10) to take the following form

$$V(\alpha,\beta)=\int_{\beta}^{\alpha}u_\alpha(\gamma,x(\beta))d\gamma+\int_{0}^{\beta}u_\alpha(\gamma,x(\gamma))d\gamma. \tag{6.14}$$

Therefore,

$$V(\alpha,\beta)-V(\alpha,\alpha)=\int_{\beta}^{\alpha}(u_\alpha(\gamma,x(\beta))-u_\alpha(\gamma,x(\gamma))d\gamma. \tag{6.15}$$

Suppose that $\alpha\geq\beta$ then $\gamma\geq\beta$ over the integration range, since $x(\cdot)$ is increasing $x(\beta)\leq x(\gamma)$ and by the single crossing property $u_\alpha(\gamma,x(\beta))\leq u_\alpha(\gamma,x(\gamma))$ for all $\gamma\in[\beta,\alpha]$, therefore $V(\alpha,\beta)-V(\alpha,\alpha)\leq 0$. Similar reasoning proves that $V(\alpha,\beta)-V(\alpha,\alpha)\leq 0$ in the case $\alpha\leq\beta$, which completes the proof. ■

This theorem allows us to reformulate the monopolist's problem. For this purpose define consumer's surplus by:

$$s(\alpha)=\max_x(u(\alpha,x)-t(x)). \tag{6.16}$$

Note that the consumer's surplus is a Fenchel $u-$conjugate of the tariff (I, Definition 56). According to the generalized envelope theorem (I, Theorem 167) $s(\cdot)$ is almost everywhere differentiable and

$$s'(\alpha)=u_\alpha(\alpha,x). \tag{6.17}$$

Using the definition of the consumer surplus to exclude the tariff from the monopolist's objective, the problem can be restated as:

$$\int_0^1(u(\alpha,x)-s(\alpha)-c(x))f(\alpha)d\alpha \tag{6.18}$$

$$s.t.s'(\alpha) \quad = \quad u_\alpha(\alpha,x),\ s(\alpha)\geq 0,\ x(\cdot)\text{ is increasing}$$

One might also ask whether a particular surplus function is implementable. The answer is: A surplus function is implementable if and only if the unique allocation that solves (6.17) is increasing. It is left as an exercise for a reader to show that if a surplus function is implementable the implementing tariff solves

$$t(x)=\max_\alpha(u(\alpha,x)-s(\alpha)), \tag{6.19}$$

that is the tariff is a Fenchel $u-$conjugate of the surplus.

Let us for a moment ignore the implementability constraint in problem (6.18). The problem we will end up with is called the relaxed problem. There are three ways to solve it. Before discussing them, however, I will introduce an important economic concept: *the information rent.*

6.2 The Concept of the Information Rent

Let us integrate equation (6.17) from α_1 to α_2. We will obtain

$$s(\alpha_2) - s(\alpha_1) \equiv I_{12} = \int_{\alpha_1}^{\alpha_2} u_\alpha(\alpha, x(\alpha))d\alpha. \tag{6.20}$$

Given any incentive compatible allocation $x(\cdot)$, integral (6.20) determines uniquely the information rent type α_2 earns over α_1. Note that the information rent depends only on allocation and does not depend on which incentive compatible mechanism is used to implement it.

The concept of information rent is of the central importance to the screening literature and is the key to the understanding the economic roots of the difference between the unidimensional and the multidimensional case. Intuitively, since in the unidimensional case there exists only one line connecting any two types, the possibility to define the information rents does not put any restrictions on the allocation. This makes the unidimensional model technically simple and amenable to a variety of approaches, which I will discuss in the next Section.

In the multidimensional case, however, any two points can be connected by a continuum of paths. Each of them can be used to define the information rent by a formula similar to (6.20). However, for the information rent to be a meaningful economic concept, this integral should be path independent. This puts severe restrictions on the set of implementable allocations and makes the multidimensional problem much harder then the unidimensional one. In particular, this is the main reason why the direct approach, which I describe in the next Section, has very limited applicability in the multidimensional case.

6.3 Three Approaches to the Unidimensional Relaxed Problem

Consider problem (6.18) and drop for a moment the constraint that $x(\cdot)$ is increasing. The resulting problem is called the relaxed screening problem. Three approaches to the solution of this problem are possible.

6.3.1 The Direct Approach

This approach was used by Mussa and Rosen (1978) and uses integration by parts technique. Let us evaluate

$$-\int_0^1 s(\alpha)f(\alpha)d\alpha = \int_0^1 s(\alpha)d(1-F(\alpha)) = -s(0) + \int_0^1 s'(\alpha)d(1-F(\alpha)).$$

Here $F(\cdot)$ is the cumulative distribution function corresponding to the density function $f(\cdot)$. Using the envelope condition, the monopolist's objective can be rewritten as

$$\int_0^1 (u(\alpha, x) - c(x) - \frac{1 - F(\alpha)}{f(\alpha)} u_\alpha(\alpha, x)) f(\alpha) d\alpha - s(0). \tag{6.21}$$

The profit maximization then implies that $s(0) = 0$ and x is a pointwise maximizer of the integrand. The first order condition is

$$(u_x(\alpha, x) - c'(x)) f(\alpha) - u_{\alpha x}(\alpha, x)(1 - F(\alpha)) = 0. \tag{6.22}$$

Note that for $\alpha = 1$ equation (6.22) implies

$$u_x(1, x) = c'(x) \tag{6.23}$$

that is the highest type consumes the good of the efficient quality. This property is known as *no distortion at the top.* In the next chapter we are going to investigate to what extend does this property generalize to the multidimensional case.

6.3.2 The Dual Approach

In this approach we start by solving equation (6.17) for x in terms of $\alpha, s,$ and s' and substitute the result into the monopolist's objective (6.18) to get a calculus of variations problem. For the expositional simplicity assume that

$$u(\alpha, x) = \alpha x. \tag{6.24}$$

Then

$$s'(\alpha) = x \tag{6.25}$$

and the monopolist's solves

$$\max \int_0^1 (\alpha s' - s(\alpha) - c(s')) f(\alpha) d\alpha. \tag{6.26}$$

The Euler equation (4.76) has a form

$$\frac{\partial}{\partial \alpha}([\alpha - c'(s')] f(\alpha)) = -f(\alpha). \tag{6.27}$$

Note that the Euler equation contains only the derivative of the surplus but not the surplus itself. This means that if $s^*(\alpha)$ is a solution, so is $s^*(\alpha) + C$. The constant should be adjusted to make profits as large as possible, subject to the participation constraint. Hence, $s(0) = 0$. Other end ($\alpha = 1$) is free, so the transversality condition holds at $\alpha = 1$

$$c'(s'(1)) = 1.$$

The Euler equation and the transversality condition imply

$$[\alpha - c'(x)]f(\alpha)) = 1 - F(\alpha), \tag{6.28}$$

which is exactly equation (6.22) for $u(\alpha, x) = \alpha x$.

6.3.3 The Hamiltonian Approach

Suppose we cannot find an explicit formula for x in terms of α, s, and s' from condition (6.17) Then we can consider the monopolist's relaxed problem is an optimal control problem.

$$\max \int_0^1 (u(\alpha, x) - s(\alpha) - c(x))f(\alpha)d\alpha$$

$$s.t. s'(\alpha) = u_\alpha(\alpha, x), \; s(0) = 0.$$

Here s is the state variable, x is the control, the left end is fixed ($s(0) = 0$) and the right end is free. Form a Hamiltonian:

$$H = (u(\alpha, x) - s(\alpha) - c(x))f(\alpha) + \lambda u_\alpha(\alpha, x). \tag{6.29}$$

Then the first order conditions have a form

$$\lambda'(\alpha) = -\frac{\partial H}{\partial s} = f(\alpha) \tag{6.30}$$

$$\lambda(1) = 0 \tag{6.31}$$

$$H_x = (u_x(\alpha, x) - c'(x))f(\alpha) + \lambda u_{\alpha x}(\alpha, x) = 0. \tag{6.32}$$

Here λ is the costate variable whose evolution is governed by equation (6.30). Economically, $\lambda(\alpha)$ is the marginal value for the monopolist of relaxing the local downward incentive compatibility constraint for type α. Equation (6.31) is the transversality condition that should hold at the free end. Finally, equation (6.32) is the Pontryagin's maximum principle.

Equations (6.30) and (6.31) imply that $\lambda(\alpha) = F(\alpha) - 1$. Substituting in into (6.32) results in

$$(u_x(\alpha, x) - c'(x))f(\alpha) - (1 - F(\alpha))u_{\alpha x}(\alpha, x) = 0. \tag{6.33}$$

Note that equation (6.33) coincides with equation (6.22). Moreover, they coincide with (6.28) for $u = \alpha x$. Therefore, all three approaches lead to the same solution of the relaxed problem. Hamiltonian approach also allows us to deal easily with type dependent participations constraints. All one has to do is to put a Lagrange multiplier on the constraint and add this term to the Hamiltonian. We will discuss this problem later in this chapter.

If the solution to the relaxed problem is weakly increasing it coincides with the solution of the complete problem. The sufficient conditions for this to be the case are

$$u_{\alpha\alpha x} \leq 0 \tag{6.34}$$

and

$$LR'(\alpha) \leq 0, \tag{6.35}$$

where the likelihood ratio, $LR(\alpha)$, is defined by:

$$LR(\alpha) = \frac{1 - F(\alpha)}{f(\alpha)}. \tag{6.36}$$

The last condition is known as the *monotone likelihood ratio property (MLRP)*. Indeed, it is easy to see that (6.34) and (6.36) imply that the monopolist's objective

$$(u(\alpha, x) - c(x)) - \frac{1 - F(\alpha)}{f(\alpha)} u_\alpha(\alpha, x) \tag{6.37}$$

is supermodular in (α, x). Therefore, the Monotone Maximum Theorem (I, Theorem 155) implies that its pointwise maximizer is increasing. If the utility has a form

$$u(\alpha, x) = \alpha x, \tag{6.38}$$

the solution to the relaxed problem will satisfy

$$c'(x) = v(\alpha), \tag{6.39}$$

where the *virtual type*, $v(\alpha)$ is defined by

$$v(\alpha) = \alpha - \frac{1 - F(\alpha)}{f(\alpha)}. \tag{6.40}$$

The solution will be implementable if and only if the virtual type is increasing in α. If the solution is not increasing in α it should be modified. We will discuss the necessary modifications in the next section.

6.4 Hamiltonian Approach to the Complete Problem

Though all three approaches described in the previous section work equally well for the relaxed problem, the first two do not allow to arrive at the solution of the complete problem in a regular way. Mussa and Rosen (1978) used some heuristic arguments to show that if the implementability constraint is binding the solution will have at least one (and probably several) bunches, i. e. segments $[a_i, b_i]$ on which $x(\alpha)$ is fixed on some constant level, $\overline{x}$. They also derived the conditions that should hold on the bunch. Here, we will show that the Hamiltonian approach can be used to arrive at the solution of the complete problem in a regular way. This approach was first taken by Fudenberg and Tirole (1992).

Let us consider the optimal control problem:

$$\max \int_0^1 (u(\alpha, x) - s(\alpha) - c(x)) f(\alpha) d\alpha \tag{6.41}$$

$$\begin{aligned} s.t.\ s'(\alpha) &= u_1(\alpha, x), & (6.42)\\ x'(\alpha) &= \gamma, & (6.43)\\ \gamma &\geq 0, & (6.44)\\ s(0) &= 0. & (6.45) \end{aligned}$$

Here conditions (6.43) and (6.44) incorporate the monotonicity constraint explicitly in the optimal control problem. The state variables for this problem are s and x and the control is γ. The constraint (6.44) is known as a *phase constraint.* (For a more thorough discussion of the optimal control problems with phase constraints, see Tikhomirov and Ioffe 1979). Form a Hamiltonian

$$H(\alpha, \gamma; x, s, \lambda_1, \lambda_2) = (u(\alpha, x) - s(\alpha) - c(x)) f(\alpha) + \lambda_1 u_1(\alpha, x) + \lambda_2 \gamma.$$

Here s and x are state variables, λ_1, λ_2 are costate variables, γ is the control variable. The evolution of the costate variables is governed by

$$\begin{aligned} \lambda_1'(\alpha) &= f(\alpha) & (6.46)\\ \lambda_2'(\alpha) &= -[(u_2(\alpha, x) - c'(x)) f(\alpha) + \lambda_1 u_{12}(\alpha, x)] & (6.47)\\ \lambda_1(1) &= 0,\ \lambda_2(1) = 0,\ \lambda_2(0) = 0. & (6.48) \end{aligned}$$

The Pontryagin's maximum principle states that

$$\begin{aligned} \gamma &\in \arg\max H(\alpha, \gamma; x, s, \lambda_1, \lambda_2) & (6.49)\\ & \quad s.t.(6.44). & (6.50) \end{aligned}$$

Let μ be the Lagrange multiplier for (6.44). Then the Kunh-Tucker necessary conditions imply:

$$\begin{aligned} \lambda_2 + \mu &= 0 & (6.51)\\ \mu &\geq 0,\ \mu\gamma = 0. & (6.52) \end{aligned}$$

Note that we could have obtained the system ((6.46)-(6.48), (6.51), (6.52) immediately by adding term $\mu\gamma$ term to the Hamiltonian. Equations (6.46) and (6.48) imply that that $\lambda_1(\alpha) = F(\alpha) - 1$. If constraint (6.44) does not bind then the complimentary slackness condition (6.52) implies that $\mu = 0$. Therefore, by (6.51) $\lambda_2 = 0$ and x is determined from

$$(u_2(\alpha, x) - c'(x))f(\alpha) + \lambda_1 u_{12}(\alpha, x) = 0, \tag{6.53}$$

which coincides with (6.33). Suppose (6.44) binds on some segment $[a, b]$. Then equation (6.43) implies that $x = \overline{x}$ on $[a, b]$ for some constant $\overline{x}$. From the continuity of the optimal allocation at a and b (see Chapter 9 for the proof of the continuity of the optimal allocation):

$$(u_2(a, \overline{x}) - c'(\overline{x}))f(a) - (1 - F(a))u_{12}(a, \overline{x}) = 0, \tag{6.54}$$
$$(u_2(b, \overline{x}) - c'(\overline{x}))f(b) - (1 - F(b))u_{12}(b, \overline{x}) = 0. \tag{6.55}$$

Moreover, since $\lambda_2(a) = \lambda_2(b) = 0$ we have

$$\int_a^b [(u_2(\alpha, \overline{x}) - c'(\overline{x}))f(\alpha) - (1 - F(\alpha)u_{12}(\alpha, \overline{x})]d\alpha = 0. \tag{6.56}$$

System (6.54)- (6.56) fully characterizes a bunch. Multiple bunches correspond to the multiple solutions of the system.

6.5 Type Dependent Participation Constraint

So far we have assumed that the value of the outside option is independent of type and normalized it to be zero. In this section I will allow for the value of the outside option to depend on type and assume that it is given by a continuously differentiable function $s_0(\alpha)$. I will also assume that the solution to the equation

$$u_\alpha(\alpha, x) = s_0'(\alpha) \tag{6.57}$$

is increasing in α, i. e. that the surplus function $s_0(\alpha)$ is implementable. Economically this means that $s_0(\alpha)$ is obtained from purchasing the good on some outside market. In particular, it may arise from the decision not to consume the good at all (i. e to set $x = 0$) in which case

$$s_0(\alpha) = u(\alpha, 0). \tag{6.58}$$

This situation naturally arises, for example, in the monopolistic insurance problem.

For the sake of simplicity let us assume that the monotonicity constraint is not binding. It is straightforward to add it, if necessary. This problem was first considered by Jullien (2000), who characterized the solution through a series of lemmata. Here, I will use the Hamiltonian approach to arrive at the same results.

Formally, the monopolist faces the following problem

$$\max \int_0^1 (u(\alpha, x) - s(\alpha) - c(x)) f(\alpha) d\alpha \tag{6.59}$$

$$s.t. s'(\alpha) = u_\alpha(\alpha, x), \tag{6.60}$$

$$s(\alpha) \geq s_0(\alpha). \tag{6.61}$$

First note that by Theorem 1 of Chapter 9.4 of Luenberger (1969), there exists a distribution η such that if pair $(x(\alpha), s(\alpha))$ solves (6.59)-(6.61) it also solves the unconstrained problem:

$$\max \int_0^1 (S(\alpha, x) - s + \eta(\alpha)(s - s_0(\alpha))) dF(\alpha) \tag{6.62}$$

$$s.t. s'(\alpha) = u_\alpha(\alpha, x), \tag{6.63}$$

where the social surplus $S(\cdot, \cdot)$ is defined by

$$S(\alpha, x) = u(\alpha, x) - c(x). \tag{6.64}$$

Moreover, $\eta(\cdot)$ satisfies

$$\eta(\alpha) \geq 0, \;\; \eta(\alpha)(s(\alpha) - s_0(\alpha)) = 0. \tag{6.65}$$

Form a Hamiltonian

$$H = (u(\alpha, x) - s(\alpha) - c(x)) f(\alpha) + \lambda u_\alpha(\alpha, x) + \eta(\alpha)(s - s_0). \tag{6.66}$$

Then the first order conditions imply

$$\left\{ \begin{array}{c} \lambda'(\alpha) = f(\alpha) - \eta(\alpha) \\ s'(\alpha) = u_\alpha(\alpha, x) \\ (u_x(\alpha, x) - c'(x)) f(\alpha) + \lambda(\alpha) u_{\alpha x}(\alpha, x) = 0 \\ \eta(\alpha)(s(\alpha) - s_0(\alpha)) = 0 \\ \lambda(0) = \lambda(1) = 0. \\ \eta(\alpha) \geq 0, \;\; s(\alpha) \geq s_0(\alpha). \end{array} \right. \tag{6.67}$$

Integrating the first equation in system (6.67) from zero to one and using the fifth equation one obtains

$$\int_0^1 \eta(\alpha) d\alpha = \int_0^1 f(\alpha) d\alpha = 1, \tag{6.68}$$

therefore $\eta(\cdot)$ is a probability distribution on $[0, 1]$. Introduce $\zeta(\cdot)$ by

$$\zeta(\alpha) = \int_0^\alpha \eta(\theta) d\theta. \tag{6.69}$$

Then following the same steps as for the problem with the type independent participation constraint one obtains

$$(u_x(\alpha, x) - c'(x)) f(\alpha) + (F(\alpha) - \zeta(\alpha)) u_{\alpha x}(\alpha, x) = 0. \tag{6.70}$$

Note that if the participation constraint (6.61) is not binding on some open interval (a, b) then $\zeta(\cdot)$ is constant on it, since $\zeta'(\alpha) = \eta(\alpha) = 0$. This is exactly the characterization obtained by Jullien (2000).

6.6 Random Participation Constraint

So far I assumed that the participation constraint is determined by the consumer type. In this section I am going to discuss the case, when the participation constraint is random. One way to look at this problem is to use the general techniques for multidimensional screening models, assuming the value of the outside option is the second dimension of type. Rochet and Stole (2001), however, proposed a simpler approach that essentially embeds the problem into the unidimensional framework. I will describe their method here.

Assume that the value of outside option, ξ, is private information of the consumer and variables α and ξ are jointly distributed according to a strictly positive density function $f(\alpha, \xi)$ on $(0, 1) \times (0, 1)$. Define

$$M(s, \alpha) = \int_0^s f(\alpha, \xi) d\xi. \tag{6.71}$$

$M(s(\alpha), \alpha)$ gives the probability that a consumer with type α will participate in the mechanism if a surplus function $s(\cdot)$ is implemented. Then the monopolist's problem becomes

$$\max_\gamma \int_0^1 [u(\alpha, x) - s - c(x)] M(s(\alpha), \alpha) d\alpha \tag{6.72}$$

$$s.t. s'(\alpha) = u_\alpha(\alpha, x), \; x'(\alpha) = \gamma, \; \gamma \geq 0. \tag{6.73}$$

This is an optimal control problem similar to (6.41)-(6.45). Note, however, that we cannot conclude here that $s(0) = 0$, therefore both ends are free and the transversality conditions should hold on each of them.

6.7 Examples

In this section I am going to solve several examples, using the techniques introduced in this chapter.

Example 173 *Assume that a consumer's utility has a form*

$$u(\alpha, x) = \alpha x \tag{6.74}$$

where α is distributed uniformly on $(0,1)$ and the monopolist's cost is given by

$$c(x) = \frac{x^2}{2}. \tag{6.75}$$

and the value of outside option is type independent and normalized to be zero. Consider first the relaxed problem. The virtual type

$$v(\alpha) = \alpha - \frac{1 - F(\alpha)}{f(\alpha)} = 2\alpha - 1 \tag{6.76}$$

and the optimal allocation is given by

$$c'(x) = x(\alpha) = \begin{cases} v(\alpha), & \alpha \geq 1/2 \\ 0, & \alpha \leq 1/2 \end{cases}. \tag{6.77}$$

This allocation is increasing in α and therefore implementable. To find the implementing tariff, first integrate the envelope condition

$$s'(\alpha) = x \tag{6.78}$$

to find

$$s(\alpha) = \begin{cases} \alpha^2 - \alpha + c, & \alpha \geq 1/2 \\ c - 1/4, & \alpha \leq 1/2 \end{cases}, \tag{6.79}$$

where constant c is found from $s(0) = 0$, which implies $c = 1/4$. Therefore,

$$t(\alpha) = \alpha x(\alpha) - s(\alpha) = \alpha^2 - \frac{1}{4}. \tag{6.80}$$

Using equation (6.77) to solve for α in terms of x one finally obtains

$$t(x) = \frac{1}{4}(x^2 + 2x). \tag{6.81}$$

Example 174 *Assume that a consumer's utility has a form*

$$u(\alpha, x) = \alpha x \tag{6.82}$$

where α is distributed uniformly on $(0,1)$ and the monopolist's cost is given by

$$c(x) = \frac{x^2}{2}. \tag{6.83}$$

and the value of outside option is type dependent given by

$$s_0(\alpha) = \alpha - \frac{3}{8}. \tag{6.84}$$

The optimal allocation is characterized by the following system

$$\left\{ \begin{array}{c} \lambda'(\alpha) = 1 - \eta(\alpha) \\ s'(\alpha) = x \\ (\alpha - x) + \lambda(\alpha) = 0 \\ \eta(\alpha)(s(\alpha) - s_0(\alpha)) = 0 \\ \lambda(0) = \lambda(1) = 0. \\ \eta(\alpha) \geq 0, \ s(\alpha) \geq s_0(\alpha). \end{array} \right. \tag{6.85}$$

Let us for a moment assume that the participation constraint does not bind. Then the solution is the same as in the previous example. In particular,

$$s(\alpha) = \alpha^2 - \alpha + \frac{1}{4}. \tag{6.86}$$

However, it is straightforward to see that $s(1) < s_0(1)$*, therefore the participation constraint should bind on* $(\alpha^*, 1)$ *for some* $\alpha^* \in [0, 1)$*. For* $\alpha > \alpha^*$ *we get*

$$x = s_0'(\alpha) = 1, \tag{6.87}$$

$$\lambda(\alpha) = x - \alpha = 1 - \alpha, \tag{6.88}$$

$$\eta(\alpha) = 1 - \lambda'(\alpha) = 2, \tag{6.89}$$

$$\zeta(\alpha) = \int_0^\alpha \eta(\theta) d\theta = 2\alpha, \tag{6.90}$$

$$s(\alpha) = s_0(\alpha) = \alpha - \frac{3}{8}. \tag{6.91}$$

For $\alpha < \alpha^*$ *on the other hand*

$$x = \max(2\alpha - \zeta, 0), \tag{6.92}$$

$$\zeta = 2\alpha^*, \tag{6.93}$$

$$s = \alpha^2 - \zeta\alpha + \frac{\zeta^2}{4}, \ \textit{for } \alpha \geq \zeta/2. \tag{6.94}$$

From continuity of the allocation and the surplus at $\alpha = \alpha^*$ *we find*

$$\zeta = \frac{1}{4}, \alpha^* = \frac{1}{8} \tag{6.95}$$

and the optimal allocation is

$$x(\alpha) = \left\{ \begin{array}{c} 0, \ \textit{if } \alpha \leq 1/8 \\ 2\alpha - 1/4, \ \textit{if } 1/8 < \alpha \leq 5/8 \\ 1, \ \textit{if } 5/8 < \alpha \leq 1. \end{array} \right. \tag{6.96}$$

The implementing tariff is given by

$$t(x) = \begin{cases} (2x^2 + x)/8, & \text{for } x \leq 1 \\ x - 5/8, & \text{for } x > 1. \end{cases} \tag{6.97}$$

Example 175 *Assume that a consumer's utility has a form*

$$u(\alpha, x) = \alpha x \tag{6.98}$$

where α is distributed uniformly on $(1, 2)$, the monopolist's cost is given by

$$c(x) = \frac{x^2}{2}. \tag{6.99}$$

and the value of outside option is type dependent given by

$$s_0(\alpha) = 0. \tag{6.100}$$

Though the value of the outside option is constant here, for the illustrative purposes we will treat this problem as a problem with a type dependent participation constraint. The first order conditions for the relaxed problem are

$$\begin{cases} s'(\alpha) = x(\alpha) \\ \lambda'(\alpha) = 1 - \eta(\alpha) \\ \alpha + \lambda = x, \; x \geq 0 \\ \eta \geq 0, \; s \geq 0, \; \eta s = 0 \\ \lambda(1) = \lambda(2) = 0 \\ \int_1^2 \eta(\alpha) d\alpha = 1 \end{cases} . \tag{6.101}$$

We will see below that the solution to the relaxed problem in this case is implementable. Note that types in close to the top type participate in the mechanism, therefore for α sufficiently close to two

$$\eta(\alpha) = 0, \; \lambda(\alpha) = \alpha - 2, \; x(\alpha) = 2\alpha - 2. \tag{6.102}$$

Note that this allocation is implementable and the corresponding surplus, found by integrating the envelope condition is

$$s(\alpha) = (\alpha - 1)^2 + c, \tag{6.103}$$

for some constant c. If $c > 0$ then the participation constraint is slack for all types in $(1, 2)$. Therefore, it is never optimal for the monopolist to choose $c > 0$ and we will write below $c = -a^2$ for some constant $a \geq 0$. In this case the participation region is given by

$$\Omega_p = (1 + a, 2) \tag{6.104}$$

and the exclusion region

$$\Omega_e = (1, 1+a). \tag{6.105}$$

In the exclusion region

$$x = 0, \;\; \lambda = -\alpha, \tag{6.106}$$

but since $\lambda(1) = 0$ one can conclude that

$$\lambda(\alpha) = -\alpha + \theta(1-\alpha), \tag{6.107}$$

where $\theta(\cdot)$ is the Heviside step function, defined in (I, Example 126). Therefore,

$$\eta(\alpha) = 1 - \lambda'(\alpha) = 2 + \delta(1-\alpha) \tag{6.108}$$

for $\alpha \in (1, 1+a)$ and

$$\eta(\alpha) = 0 \tag{6.109}$$

for $\alpha \in (1+a, 2)$. Hence, one obtains

$$\int_1^2 \eta(\alpha) d\alpha = 1 + 2a = 1, \tag{6.110}$$

which implies $a = 0$. Therefore, all the types participate in the mechanism, the optimal allocation is

$$x(\alpha) = 2\alpha - 2, \tag{6.111}$$

and the implementing tariff is

$$t(x) = \frac{x^2}{4} + x. \tag{6.112}$$

The Lagrange multipliers that support this allocation are

$$\lambda(\alpha) = \alpha - 2 + \theta(1-\alpha) \tag{6.113}$$
$$\eta(\alpha) = \delta(1-\alpha). \tag{6.114}$$

The purpose of this example is to show that the Lagrange multipliers in the screening models can be distributions rather than functions even if all the data of the problem are smooth.

Example 176 *This example was first considered by Bardsley and Basov (2003). Assume that the research institutions are competing for projects. Each institution is characterized by a success probability, α, and the value of outside option, ξ, which are are jointly distributed according to a strictly positive density function $f(\alpha, \xi)$ on $(0,1) \times (0,1)$. The public value of the project is one. Assume that the government offers a payment schedule*

$$\pi = \pi(p), \tag{6.115}$$

where p is the probability of success that an institution announces in the grant application and π is the size of the grant. Assume also that an institution faces a cost of lying

$$c(p-\alpha) = \frac{1}{2}[\max(0,(p-\alpha))]^2, \tag{6.116}$$

It can be justified by assuming that the inflated success probability might become detected by academic peers, in which case the institution will loose its reputation. Another possibility is that the cost of lying is purely psychological. Therefore, given the schedule (6.115) an institution solves

$$\max_p(\pi(p) - \frac{1}{2}[\max(0,(p-\alpha))]^2) \tag{6.117}$$

Applying the envelope theorem to (5) one obtains

$$s'(\alpha) = \max(0, p-\alpha). \tag{6.118}$$

The government seeks to maximize the expected value of the projects minus financing costs, that is it solves

$$\max \int_0^1\int_0^1 (\alpha - \pi(p))f(\alpha,\xi)d\alpha d\xi, \tag{6.119}$$

$$\begin{aligned} s.t.\ s'(\alpha) &= \max(0,p-\alpha), 0 \le p \le 1, s(0) = 0, & (6.120)\\ p'(\alpha) &\ge 0. & (6.121)\end{aligned}$$

Let us introduce $M(\cdot,\cdot)$ by (6.71) and concentrate for a moment on the relaxed problem. Then the government's problem becomes

$$\max \int_0^1 [\alpha - s - \frac{1}{2}[\max(0,(p-\alpha))]^2)Md\alpha, \tag{6.122}$$

$$s.t.\ s'(\alpha) = \max(0,\ p-\alpha),\ 0 \le p \le 1. \tag{6.123}$$

Following the usual procedure for the optimal control problems with phase constraints write the Hamiltonian in the form

$$H = [\alpha - s - c(p-\alpha)]M(\alpha,s) + \lambda(p-\alpha) - \mu(p-1). \tag{6.124}$$

Then the first order conditions are

$$\begin{aligned} \lambda'(\alpha) &= M + (s + c(p-\alpha) - \alpha)M_s & (6.125)\\ c'(p-\alpha)M(\alpha,s) - \lambda + \mu &= 0 & (6.126)\\ s'(\alpha) &= c'(p-\alpha) & (6.127)\\ \mu &\ge 0,\ p \le 1,\ \mu(p-1) = 0 & (6.128)\\ \lambda(0) &= 0,\ \lambda(1) = 0 & (6.129)\end{aligned}$$

The steps for solving the problem are following:
Step 1. *Since $p(\alpha)$ is an increasing function of α, the constraint $p \leq 1$ binds on $[\alpha^*, 1]$ for some $\alpha^* \in [0,1]$. On this interval (6.127) implies*

$$s(\alpha) = s_0 - c(1 - \alpha).$$

Step 2. *Using (6.125), (6.126), and the expression for $u(\alpha)$ obtained at step one, it is possible to solve for a derivative of $\mu(\cdot)$:*

$$\mu'(\alpha) = W(\alpha)$$

Equations (6.126) and (6.129) imply $\mu(1) = 0$. Hence, one can determine μ as a function of α and parameter $s_0 : \mu(\alpha, s_0)$.
Step 3. *Define $s_0(\alpha^*)$ to be a solution to the equation*

$$\mu(\alpha^*, s_0) = 0.$$

Step 4. *Equations (6.125) and (6.126) with $\mu = 0$ imply*

$$s' M_\alpha = (1 - s'')M + (s + \frac{s'^2}{2} - \alpha)M_s \tag{6.130}$$

The continuous differentiability of $s(\cdot)$ at α^* implies

$$\begin{aligned} s(\alpha^*) &= s_0(\alpha^*) - c(1 - \alpha^*) \\ s'(\alpha^*) &= c'(1 - \alpha^*). \end{aligned}$$

Step 5. *Fix a value of α^* and solve the Cauchy problem obtained at Step 4. Check the transversality condition $\lambda(0) = 0$ and the inequality $\mu(\alpha) \geq 0$ at $[\alpha, 1]$ (the necessary condition for this is $\mu'(1) \leq 0$). Adjust α^* until these conditions are satisfied.*

Let us carry out all the steps for the case, where α and ξ are distributed independently and uniformly on the unit square. In this case $M(\alpha, s) = s$.

$$s''s + \frac{s'^2}{2} = 2s - \alpha - \mu'(\alpha), \tag{6.131}$$

For $\alpha \in [\alpha^*, 1]$

$$\begin{aligned} s &= s_0 + \alpha - \frac{\alpha^2}{2} && (6.132) \\ \mu'(\alpha) &= 3s_0 - \frac{1}{2} + 3\alpha - 2\alpha^2 && (6.133) \\ \mu(1) &= 0. && (6.134) \end{aligned}$$

Hence,

$$\mu(\alpha) = (3s_0 - \frac{1}{2})\alpha + \frac{3}{2}\alpha^2 - \frac{2}{3}\alpha^3 - (3s_0 + \frac{1}{3}) \tag{6.135}$$

For $\alpha \in [0, \alpha^*]$ we solve Cauchy problem

$$s''s + \frac{s'^2}{2} = 2s - \alpha \tag{6.136}$$

$$s(\alpha^*) = -\frac{5}{18}\alpha^{*2} + \frac{13}{18}\alpha^* - \frac{1}{9}, \; s'(\alpha^*) = 1 - \alpha^*. \tag{6.137}$$

The transversality condition in this case can be reduced to

$$s'(0, \alpha^*) = 0. \tag{6.138}$$

Solving the system numerically, we obtained $\alpha = 0.979$, $s_0 = -0.170$. It is easy to check that the solution is implementable. One can verify that $\mu(\alpha)$ is positive on $(0.979, 1)$. Note that $s(0) = 0.132$, hence the lowest type obtains positive surplus. Also note, that $s(\alpha) > \alpha$ for $\alpha < 0.149$. Since the participation region is given by

$$\{(\alpha, \xi) \in [0,1] \times [0,1] : \xi \leq s(\alpha)\} \tag{6.139}$$

this implies that too much difficult and too few easy projects are financed.

6.8 Exercises

1. Prove that if the surplus is implementable the implementing tariff satisfies equation (6.19).
2. Assume that a consumer's utility is

$$u(\alpha, x) = \alpha^2 \sqrt{x}. \tag{6.140}$$

Are the following surpluses implementable (if yes, find the implementing tariff):
  a). $s(\alpha) = \alpha^{3/2}$
  b). $s(\alpha) = \alpha^3$
  c). $s(\alpha) = \alpha^2$
  d). $s(\alpha) = \alpha^2 + \alpha^3 \ln \alpha$.
3. Let

$$u(\alpha, x) = \alpha x, \tag{6.141}$$

$$c(x) = \frac{x^2}{2}, \tag{6.142}$$

and

$$f(\alpha) = \frac{e}{e-1} \exp(-\alpha). \tag{6.143}$$

Find the optimal allocation and implementing tariff.

6.9 Bibliographic Notes

The topic of imperfect information first attracted researches in the middle of the seventies. Probably the first paper in the area is Mirrlees (1971). Some of the early papers (e. g. Adams and Yellen, 1976) used a finite number of types. For an excellent review of such models and the main result in that area, the so called constraint reduction theorem, one can consult the still unpublished but available on the web notes of Stole (2000). From Mussa and Rosen (1978) on the interest firmly shifted to the models with a continuum of types. Similar models, arising from different economic situations, were considered by Mirman and Sibley (1980), Baron and Myerson (1982), Sappington (1983), and Maskin and Riley (1984) among others. By now the economics of the basic model is well understood and the material of the first four sections of this chapter found its way into numerous textbooks (see, for example, Fudenberg and Tirole 1992). It is included here only for the completeness of exposition.

The first example of the model with type dependent participation constraint is the countervailing incentives model of Lewis and Sappington (1989). A general treatment of such models is given by Jullien (2000). The investigation of the random participation constraint model is given in an unpublished but available on the web paper by Rochet and Stole (2001).

The main contribution of this work to the area of the unidimensional screening models is methodological. Here I derive all the results using the Hamiltonian approach, while the previous papers usually used some case specific arguments.

7 The Multidimensional Screening Model

In this chapter we will discuss the multidimensional screening problem, that is the problem when the private information of the consumer cannot be captured in one characteristic. The general formulation of the problem of multidimensional screening is due to Armstrong (1996) and Wilson (1993), and goes as follows. Consider a monopoly producing a good with n quality characteristics. Each consumer is interested in buying at most one unit of the monopolist's good and her preferences over the quality can be parameterized by an $m-$dimensional vector.

More formally, assume a monopolist who faces a continuum of consumers produces a good of with n quality dimensions, which can be captured by a vector $\mathbf{x} \in R^n$. For example, if the monopolist produces cars, then x_1 can be the maximal speed, x_2- the power of the air conditioning, x_3- the engine efficiency, etc. The cost of production is given by a twice differentiable function $c(\cdot)$, which is convex in the quality and linear in the number of goods produced. Each consumer is interested in consuming at most one unit of the good and has a utility

$$u(\boldsymbol{\alpha}, \mathbf{x}) - t, \tag{7.1}$$

where $\alpha \in R^m$ is her unobservable type distributed on an open, bounded, convex set $\Omega \subset R^m$ according to a strictly positive, continuous density function $f(\cdot)$, t is the amount of money transferred to the monopolist, and $u(\alpha,x)$ is a continuous function, strictly increasing in both arguments. Consumers have an outside option of value $s_0(\alpha)$. The monopolist is looking for a mechanism that would maximize her profits.

Reasoning in the same way as in the unidimensional case, one can without loss of generality assume that the monopolist simply announces a non-linear tariff $t(\cdot)$. The amount $t(\mathbf{x})$ determines how much a consumer has to pay fro a good with quality characteristics $\mathbf{x}$.

The above consideration can be summarized by the following model. The monopolist selects a continuos $t : R^n_+ \to R$ to solve

$$\max_{t(\cdot)} \int_{\Omega} (t(\mathbf{x}(\boldsymbol{\alpha})) - c(\mathbf{x}(\boldsymbol{\alpha})))f(\boldsymbol{\alpha})d\boldsymbol{\alpha}, \tag{7.2}$$

where $c(\mathbf{x})$ is cost of producing a good with quality $\mathbf{x}$ and $\mathbf{x}(\alpha)$ satisfies

$$\mathbf{x}(\boldsymbol{\alpha}) \in \arg\max(u(\boldsymbol{\alpha}, \mathbf{x}) - t(\mathbf{x})) \tag{7.3}$$

$$\max(u(\boldsymbol{\alpha}, \mathbf{x}) - t(\mathbf{x})) \geq s_0(\boldsymbol{\alpha}). \tag{7.4}$$

In this chapter I discuss the monopolist's problem (7.2)-(7.4) in the general case and emphasize the essential economic and technical differences between the unidimensional and the multidimensional cases.

I start with discussing the genericity of exclusion. While in the unidimensional case it was possible to make assumptions on the distribution of types, the production cost, and the utilities that ensured that all types are served in the equilibrium, this is impossible in multidimensional case. In fact, under some reasonable assumptions on the utilities and the type set, exclusion is generic. This result was first obtained by Armstrong (1996) and signifies the first important qualitative difference between the unidimensional and the multidimensional case.

Then I move on to discuss the issue of implementability in the multidimensional case. I start with discussing the implementability of the surplus function. The necessary and sufficient condition, its generalized convexity, is a direct generalization of the requirement that in the unidimensional case the implementable allocation should be weakly increasing in type. It is captured by the Carlier's (2002) Implementability Criterion. In a nutshell, it says that if an allocation rule and at implementable by a *particular* tariff, the generalized Fenchel conjugate of the surplus, they are not implementable by *any* tariff. Hence, one can constructively check implementability given the surplus and the allocation rule.[1] Economically, it means that if a surplus is implementable the corresponding allocation should assign the good to the consumer who value it most given the allocation.

Another requirement for implementability in the multidimensional case, the requirement that a certain vector field should be conservative, has no unidimensional analog. It has its roots in the fact that the local incentive compatibility, which did not put any restrictions on the set of imple-

[1] Implementability criterion developed by Rochet (1987) does not allow for such a check.

mentable allocations for $m = 1$, becomes restrictive for $m > 1$. To understand why, recall that in the discrete unidimensional model (assuming single-crossing) it is always the downward slopping incentive compatibility between a type and its immediate neighbor to the left that binds (Stole, 2000), while in the multidimensional case the direction in which the local incentive compatibility binds is uncertain. We will see that economics of this phenomenon is closely related to the requirement of existence of a well-defined information rents and mathematically reduces to the statement that a certain vector field is conservative.

Then I will study the relaxed problem and discuss direct and dual approaches in multidimensional case and their limitations. The main limitation of the direct approach comes from the fact that it does not allow to accommodate naturally the requirement of existence of well defined information rents. Dual approach is limited by the requirement that one has to solve for instruments explicitly in terms of types and the first derivatives of the surplus and breaks completely if $m > n$. The Hamiltonian approach, however, can be applied to a general relaxed problem. Moreover, if the utility is linear in types it can be applied to characterize the solution of the complete problem. We have already seen that the Hamiltonian approach allows for a unified presentation of all known results in the unidimensional case. However, all while results in the unidimensional case can be derived using different reasoning (though not in such an elegant way), in the multidimensional case the Hamiltonian approach becomes indispensable.

7.1 The Genericity of Exclusion

Let us assume for the duration of this section that the outside option is type independent and normalize it to be zero, i. e. $s_0(\boldsymbol{\alpha}) = 0$. In the unidimensional case one can impose some relatively weak requirements on the distribution of types, the consumers' utility and the monopolist's technology that will guarantee that all consumers all served in equilibrium. For example, if the single-crossing property holds, the utility is concave in x, thrice differentiable with

$$\frac{\partial^3 u}{\partial \alpha^2 \partial x} \leq 0, \ \frac{\partial^3 u}{\partial \alpha \partial x^2} \geq 0, \tag{7.5}$$

the cost is increasing, differentiable, and strictly convex, the hazard rate

$$h(\alpha) = \frac{f(\alpha)}{1 - F(\alpha)} \tag{7.6}$$

is increasing in α, and if

$$u_x(\alpha_*, 0) - c'(0) - \frac{u_{\alpha x}(\alpha_*, 0)}{f(\alpha_*)} \geq 0, \tag{7.7}$$

where $\alpha_* = \inf \Omega$, then all types are served in equilibrium. A natural question is whether it is possible to find similar conditions in the multidimensional case.

This question was first considered by Armstrong (1996) and was answered in negative, provided that the utility is convex, increasing, and homogenous of degree one in types and Ω is a strictly convex set. The intuition behind this result is rather simple and goes as follows: Suppose tariff $t(\cdot)$ is such that all consumers are served in equilibrium and let us denote the corresponding surplus $s(\cdot)$. First, note that there exists $\boldsymbol{\alpha}_0 \in \Omega \cup \partial\Omega$ such that $s(\boldsymbol{\alpha}_0) = 0$. Indeed, otherwise

$$\inf_{\boldsymbol{\alpha}\in\Omega} s(\boldsymbol{\alpha}) = a > 0 \tag{7.8}$$

and tariff $t^*(\mathbf{x}) = t(\mathbf{x}) + a$ implements the same allocation and therefore insures higher profits. Recall that

$$s(\boldsymbol{\alpha}) = \max_{\mathbf{x}\in R^n_+} (u(\boldsymbol{\alpha}, \mathbf{x}) - t(\mathbf{x})). \tag{7.9}$$

Therefore, provided that $u(\cdot, \mathbf{x})$ is convex, so is $s(\cdot)$ as an upper envelope of the system of convex functions. Now consider a new tariff $\tau(\cdot)$ defined by

$$\tau(\mathbf{x}) = t(\mathbf{x}) + \varepsilon, \tag{7.10}$$

for some constant $\varepsilon > 0$. Note that conditional on the participation the consumers will not change the bundle they purchase. Therefore, the monopolist will earn additionally $\varepsilon(1-\lambda)$ and loose $\lambda\bar{t}$, where λ is the measure of the consumers which dropped out after the increase in the tariff and $\bar{t}$ the average amount they paid. I am going to argue that

$$\lambda = O(\varepsilon^m), \tag{7.11}$$

therefore if $m > 1$ for sufficiently small ε gains will always outweighed the losses. Therefore, the full participation is never optimal.

To see why equation (7.11) holds, note that λ is the Lebesgue measure of the following set

$$D = \{\boldsymbol{\alpha} \in \Omega : s(\boldsymbol{\alpha}) \leq \varepsilon\}. \tag{7.12}$$

Note that $\boldsymbol{\alpha}_0 \in D$. Let us assume that $s(\cdot)$ is twice differentiable in a neighborhood of $\boldsymbol{\alpha}_0$, then one can write

$$s(\boldsymbol{\alpha}) = s(\boldsymbol{\alpha}_0) + (\boldsymbol{\alpha} - \boldsymbol{\alpha}_0)\nabla s(\boldsymbol{\alpha}_0) + \frac{\partial^2 s(\boldsymbol{\xi})}{\partial\alpha_i\partial\alpha_j}(\alpha_i - \alpha_{0i})(\alpha_j - \alpha_{0j}) \tag{7.13}$$

for some $\boldsymbol{\xi}$ with $\zeta_j \in (\alpha_{0j}, \alpha_j)$. Since $s(\cdot)$ is convex the matrix

$$A = \frac{\partial^2 s(\boldsymbol{\alpha}_0)}{\partial\alpha_i\partial\alpha_j} \tag{7.14}$$

is positively semidefinite. Assume that it is positive definite, and for $\boldsymbol{\alpha}$ close enough to $\boldsymbol{\alpha}_0$ replace $\boldsymbol{\xi}$ with $\boldsymbol{\alpha}_0$. Then equation (7.13) implies that set D is an ellipsoid with axes proportional to ε, therefore

$$\lambda = mes(D) = O(\varepsilon^m). \tag{7.15}$$

After presenting the intuition I will give a formulation and a rigorous proof of this result. To do this I need the following definition. Recall that a set Ω is called strictly convex if for any $\alpha, \beta \in \Omega$ and any $\lambda \in (0, 1)$

$$\gamma = \lambda\alpha + (1 - \lambda)\beta \in Int(\Omega), \tag{7.16}$$

i. e. there exists neighborhood $U(\gamma)$ of γ such that $U(\gamma) \subset \Omega$. The main result of the section is given by the following theorem:

Theorem 177 *(Armstrong, 1996) Let $u(\cdot, \mathbf{x})$ be convex, increasing, and homogenous of degree one, and assume that $\Omega \subset R^m$ $(m \geq 2)$ be a strictly convex open set of a full measure. Let tariff $t(\cdot)$ solve (7.2)-(7.4) with $s_0(\boldsymbol{\alpha}) = 0$ and define $s(\cdot)$ to be the corresponding surplus and $\mathbf{x}(\cdot)$ the corresponding allocation. Then $mes(E) > 0$, where $mes(\cdot)$ is the Lebesgue measure and*

$$E = \{\boldsymbol{\alpha} \in \Omega : s(\boldsymbol{\alpha}) = 0\} \tag{7.17}$$

is the set of types excluded under tariff $t(\cdot)$.

Proof. Assume to the contrary, that $s(\boldsymbol{\alpha}) > 0$ for almost all $\boldsymbol{\alpha} \in \Omega$. Extend $s(\cdot)$ on $\partial\Omega$ by continuity, take some $\varepsilon > 0$ and define

$$\Omega(\varepsilon) = \{\boldsymbol{\alpha} \in \overline{\Omega} : s(\boldsymbol{\alpha}) \leq \varepsilon\}, \tag{7.18}$$

where $\overline{\Omega}$ is the closure of the set Ω. Note that since Ω is bounded, $\overline{\Omega}$ is compact. Recall that

$$s(\boldsymbol{\alpha}) = \max_{\mathbf{x} \in R_+^n} (u(\boldsymbol{\alpha}, \mathbf{x}) - t(\mathbf{x})) \tag{7.19}$$

and since $u(\cdot, \mathbf{x})$ is continuous, convex and increasing, so is $s(\cdot)$. Therefore, $\Omega(\boldsymbol{\varepsilon})$ is a convex, compact set. Moreover, $\Omega(0) \neq \varnothing$, since otherwise

$$\inf_{\boldsymbol{\alpha} \in \Omega} s(\boldsymbol{\alpha}) = a > 0 \tag{7.20}$$

and tariff $t^*(\mathbf{x}) = t(\mathbf{x}) + a$ will earn for the monopolist higher profits. Let us prove that $\Omega(0)$ is a singleton. For this assume to the contrary that $\exists \boldsymbol{\alpha}, \boldsymbol{\beta} \in \overline{\Omega}$ such that $\boldsymbol{\alpha} \neq \boldsymbol{\beta}$ and $\boldsymbol{\alpha}$, $\boldsymbol{\beta} \in \Omega(0)$ and let

$$\gamma = \frac{1}{2}\boldsymbol{\alpha} + \frac{1}{2}\boldsymbol{\beta}. \tag{7.21}$$

Then $\gamma \in Int(\Omega)$ and therefore the set

$$A = \{\boldsymbol{\chi} \in \Omega : \chi \leq \gamma\} \tag{7.22}$$

has positive Lebesgue measure. On the other hand, since $s(\cdot)$ is increasing $A \subset \Omega(0)$. Therefore, since we assumed that almost all types receive positive surplus, $\Omega(0)$ must be a singleton, $\Omega(0) = \{\mathbf{a}\}$.

Let

$$X = \{B : B \subset \Omega,\ B \text{ is compact}\}. \tag{7.23}$$

Let us endow X with the Hausdorff metric $h(\cdot, \cdot)$ (see, I, 5.150). Then $(X, h))$ is a metric space and mapping $\Omega(\cdot) : (R, d) \to (X, h)$ given by (7.18) is a continuous mapping. Recall that if a sequence of sets converges to a singleton in the Hausdorff metric the corresponding sequences of volumes and surface areas converges to zero (see I, Lemmata 166 and 168). Therefore,

$$\lim_{\varepsilon \to 0} V(\Omega(\varepsilon)) = \lim_{\varepsilon \to 0} S(\Omega(\varepsilon)) = 0. \tag{7.24}$$

Profits the monopolist obtains from consumers in set $\Omega(\varepsilon)$ are given by

$$\pi(\varepsilon) = \int\limits_{\Omega(\varepsilon)} (u(\boldsymbol{\alpha}, \mathbf{x}(\boldsymbol{\alpha})) - s(\boldsymbol{\alpha}) - c(\mathbf{x}))f(\boldsymbol{\alpha})\mathbf{d}\boldsymbol{\alpha} \leq \int\limits_{\Omega(\varepsilon)} (u(\boldsymbol{\alpha}, \mathbf{x}(\boldsymbol{\alpha}))f(\boldsymbol{\alpha})\mathbf{d}\boldsymbol{\alpha}. \tag{7.25}$$

Since $u(\cdot, \mathbf{x})$ is homogeneous of degree one

$$u(\boldsymbol{\alpha}, \mathbf{x}(\boldsymbol{\alpha})) = \boldsymbol{\alpha} \cdot \boldsymbol{\nabla}_{\alpha} u(\boldsymbol{\alpha}, \mathbf{x}(\boldsymbol{\alpha})). \tag{7.26}$$

Now, by the envelope theorem $\boldsymbol{\nabla}_{\alpha} u(\boldsymbol{\alpha}, \mathbf{x}(\boldsymbol{\alpha})) = \boldsymbol{\nabla} s(\boldsymbol{\alpha})$ and

$$\pi(\varepsilon) \leq \int\limits_{\Omega(\varepsilon)} (\boldsymbol{\alpha} \cdot \boldsymbol{\nabla} s(\boldsymbol{\alpha}))f(\boldsymbol{\alpha})\mathbf{d}\boldsymbol{\alpha}. \tag{7.27}$$

Using the Gauss-Ostragradsky Theorem (I, Theorem 14) and the fact that

$$div(\boldsymbol{\alpha} s(\boldsymbol{\alpha})f(\boldsymbol{\alpha})) = \boldsymbol{\alpha} \cdot \boldsymbol{\nabla} s(\boldsymbol{\alpha})f(\boldsymbol{\alpha}) + s(\boldsymbol{\alpha}) div(\boldsymbol{\alpha} s f(\boldsymbol{\alpha})) \tag{7.28}$$

one obtains

$$\pi(\varepsilon) \leq -\int\limits_{\Omega(\varepsilon)} s(\boldsymbol{\alpha}) div(\boldsymbol{\alpha} s f(\boldsymbol{\alpha}))\mathbf{d}\boldsymbol{\alpha} + \oint\limits_{\partial\Omega(\varepsilon)} \boldsymbol{\alpha} s(\boldsymbol{\alpha})f(\boldsymbol{\alpha})\mathbf{d}\boldsymbol{\Sigma}. \tag{7.29}$$

Applying the Mean Value Theorem (see, for example, Smirnov, 1964) one obtains

$$\pi(\varepsilon) \leq -\varepsilon K(V(\Omega(\varepsilon) + S(\Omega(\varepsilon)), \tag{7.30}$$

for some constant $K > 0$.

Now, let us consider the change in profits induced by increase in tariff by some $\varepsilon > 0$. Since consumers whose type belong to $\Omega(\varepsilon)$ will now drop out the monopolist will lose $\pi(\varepsilon)$, on the other hand she will gain on the remaining consumers at least

$$\varepsilon(1 - KV(\Omega(\varepsilon)), \tag{7.31}$$

therefore change in profit

$$\Delta\pi \geq \varepsilon(1 - (1+K)V(\Omega(\varepsilon)) - KS(\Omega(\varepsilon))) > 0 \tag{7.32}$$

for sufficiently small ε. Therefore, the initial tariff was not optimal. The contradiction proves the Theorem. ■

7.2 Generalized Convexity and Implementability

Consider problem (7.2)-(7.4). Following the same logic as in the unidimensional case let us replace global incentive compatibility constraint (7.4) by some conditions on the derivatives of the surplus functions. For this purpose, note that the envelope theorem implies that the surplus is everywhere continuous and almost everywhere differentiable and in the differentiability points the envelope conditions hold

$$\nabla s(\boldsymbol{\alpha}) = \nabla_{\boldsymbol{\alpha}} u(\boldsymbol{\alpha}, \mathbf{x}(\boldsymbol{\alpha})). \tag{7.33}$$

It also implies that the most natural space for $s(\cdot)$ to live in is $H^1(\Omega)$, which we will assume from now on.

The relaxed problem can be formulated as follows:

$$\max_{x(\cdot)} \int_{\Omega} (u(\boldsymbol{\alpha}, \mathbf{x}) - s(\boldsymbol{\alpha}) - c(\mathbf{x})) f(\boldsymbol{\alpha}) d\boldsymbol{\alpha} \tag{7.34}$$

subject to (7.33) and $s(\alpha) \geq s_0(\boldsymbol{\alpha})$. The solution to this problem need not, however, be implementable. To discuss the issue of implementability let us start with the following definition.

Definition 178 $u(\alpha, \mathbf{x})$ *is said to satisfy the generalized single-crossing (GSC) property if*

$$[\nabla_{\boldsymbol{\alpha}} u(\boldsymbol{\alpha}, \mathbf{x}_1) = \nabla_{\boldsymbol{\alpha}} u(\boldsymbol{\alpha}, \mathbf{x}_2)] \Rightarrow (\mathbf{x}_1 = \mathbf{x}_2). \tag{7.35}$$

If $m = 1$ and the Spence-Mirrlees condition holds then $u_\alpha(\alpha, \mathbf{x})$ is strictly increasing in $\mathbf{x}$, therefore the GSC holds. The following result holds:

Theorem 179 *(Carlier's Implementability Criterion) Assume $u(\alpha, \mathbf{x})$ is continuous in both arguments, continuously differentiable in α and satisfies GSC. Surplus $s(\alpha)$ is implementable if and only if it is continuous, almost everywhere differentiable, and $u-$convex. Moreover, if $s(\cdot)$ is differentiable at α, then the corresponding allocation $\mathbf{x}(\alpha)$ is the unique solution of (7.33).*

Proof. Suppose $s(\cdot)$, $x(\cdot)$ satisfies the conditions of the Theorem. Define a tariff $t(\cdot)$ by

$$t(\mathbf{x}) = \sup(u(\boldsymbol{\alpha}, \mathbf{x}) - s(\boldsymbol{\alpha})) \equiv s^*(\mathbf{x}). \tag{7.36}$$

Since $s(\cdot)$ is $u-$convex

$$s(\boldsymbol{\alpha}) = s^{**}(\boldsymbol{\alpha}) = \max(u(\boldsymbol{\alpha}, \mathbf{x}) - s^*(\mathbf{x})). \tag{7.37}$$

Therefore tariff (7.36) implements $s(\cdot)$. Moreover, by the envelope theorem the corresponding allocation, $\mathbf{h}(\cdot)$ satisfies (7.33). By GSC property $\mathbf{x}(\alpha) = \mathbf{h}(\alpha)$ at points of the differentiability of the surplus.

Now suppose that $s(\cdot)$ is implementable and $t(\cdot)$ is the implementing tariff. Then $s(\alpha) = t^*(\alpha)$, i. e. it is $u-$Fenchel conjugate of the tariff. Therefore, it is $u-$convex (see, I, Definition 59). By the envelope theorem, $s(\cdot)$ is continuos and almost everywhere differentiable. Moreover, if $s(\cdot)$ is differentiable at α, then the corresponding allocation $\mathbf{x}(\alpha)$ is the unique solution of (7.33). ■

Recall that a function is $u-$convex if and only if it equals to its generalized $u-$biconjugate (I, Theorem 63). Observe also that the only place the GSC was used was to prove that $\mathbf{x}(\cdot)$ is the *unique* allocation satisfying the envelope conditions. Using these observations, one can reformulate the above Theorem in more economically intuitive terms without requiring the GSC: *A surplus function is implementable if and only if it is implementable by the tariff equal to its u−conjugate.* Intuitively, if a surplus can be implemented at all, it can be implemented by a mechanism that allocates the good to the type who is willing to pay the most for it conditional on achieving the given surplus, and extracts all the payment consistent with it.

It is clear that if $u(\alpha, \mathbf{x})$ is linear in types the Carlier's Implementability Criterion will imply that a surplus function is implementable if an only if it is convex. This result was first obtained by Rochet (1987). Below I provide some examples of classes of implementable surpluses for particular utility functions. Let us first consider the case when the consumer's preferences are Leontieff. We will see that in this case a class of implementable allocations is rather broad. Both examples use results of Section (I, 3.3).

Example 180 *Assume*

$$u(\boldsymbol{\alpha}, \mathbf{x}) = \min_{i=1,\ldots,n} \boldsymbol{\alpha}_i \mathbf{x}_i.$$

The class of $u-$convex functions in this case is the class of lower semi-continuous convex along rays functions . Hence, any such function can be implemented as a consumer surplus by means of appropriately chosen tariff. This class is rather broad. It includes all convex functions and all functions homogenous of degree β, provided $\beta \geq 1$.

The next example describes a class of utility functions for which the implementability constraint is not binding at all.

Example 181 *Let $n = m + 1$ and*

$$u(\boldsymbol{\alpha}, \mathbf{x}) = \sum_{i=1}^{m} \alpha_i \mathbf{x}_i - \frac{1}{2} \|\boldsymbol{\alpha}\|^2 \mathbf{x}_{m+1}.$$

Then any continuous function on Ω is $u-$convex and therefore, implementable. Hence, in this case the solutions to the relaxed and to the complete problems coincide.

The generalized convexity of the surplus is a direct generalization of the monotonicity requirement for the unidimensional case. To see this for the case when

$$u(\boldsymbol{\alpha}, \mathbf{x}) = \sum_{i=1}^{n} \alpha_i \mathbf{x}_i \tag{7.38}$$

note that in this case the envelope conditions (7.33) are

$$\mathbf{x}_i = s_{\alpha_i}. \tag{7.39}$$

Convexity of the surplus implies now that $\mathbf{x}_i$ is increasing in α_i. For the general case the proof is a little bit more tedious. For details, see Basov (2002).

7.2.1 Is Bunching Robust in the Multidimensional Case?

As I discussed in Chapter 6, in the unidimensional case it is possible to come up with simple requirements on the distribution of types that will rule out bunching. Conventional wisdom, first articulated by Rochet and Chone (1998), states that bunching is robust in multidimensional case. Intuitively, the robustness of bunching comes from the conflict to serve as many consumers as possible and the requirement that the consumer surplus were convex, which is necessary for implementability if the consumers' utility is convex in types. One can observe, however, that this intuition suggests that the robustness of bunching springs from the interaction of multidimensionality with convexity of the utility in types and the convexity of the type space, i. e. multidimensionality alone is not sufficient for the robustness of bunching.

To understand better the role of multidimensionality, note that since types are just labels for the utility one can always relabel them using a diffeomorphism $h : \Omega \to \Omega'$. Let $\boldsymbol{\beta} = h(\boldsymbol{\alpha})$. Even if the utility $u(\boldsymbol{\alpha}, \mathbf{x})$ was not convex in types, one can choose the diffeomorphism in such a way that utility

$$v(\boldsymbol{\beta}, \mathbf{x}) = u(h^{-1}(\boldsymbol{\beta}), \mathbf{x}) \tag{7.40}$$

will be convex in $\boldsymbol{\beta}$. Therefore, one can view convexity of utility alone as the property of a *representation,* which is consequently not restrictive. It, however, becomes restrictive when combined with convexity of Ω, provided $m > 1$.[2] Therefore, robustness of bunching springs from interaction of three conditions: the multidimensionality of the type space, the convexity of the type space, and the convexity of the utility in types. To develop some intuition about the clash between the convexity of utility in types and convexity of Ω consider the following example.

Example 182 *Let* $\Omega = \{\boldsymbol{\alpha} \in R_+^2 : 1/4 < \alpha_1 + \alpha_2 < 1\}$ *and the utility is given by*

$$u(\boldsymbol{\alpha}, \mathbf{x}) = \sqrt{\alpha_1} x_1 + \sqrt{\alpha_2} x_2. \tag{7.41}$$

Note that while Ω *is convex, the utility (7.41) is not convex in* $\boldsymbol{\alpha}$. *Consider a change of variables:*

$$\begin{cases} \beta_1 = \sqrt{\alpha_1} \\ \beta_2 = \sqrt{\alpha_2} \end{cases}. \tag{7.42}$$

Then the utility (7.41) becomes

$$u(\boldsymbol{\beta}, \mathbf{x}) = \beta_1 x_1 + \beta_2 x_2. \tag{7.43}$$

It is now convex in types but the type space

$$\Omega' = \{\boldsymbol{\beta} \in R_+^2 : 1/2 < \beta_1^2 + \beta_2^2 < 1\} \tag{7.44}$$

is no longer convex. It is possible to show that in this case there are no diffeomorphisms of the type space that preserve its convexity and make the utility convex in the new types.

The assumptions that utility in convex in types and the type space is convex can be restated as a *joint convexity assumption*: There exists a parametrization of the consumers preferences such that the type space is convex and the utility is convex in types. The above example shows that the joint convexity assumption is restrictive in the multidimensional case. Therefore, even if bunching is robust under it, one cannot conclude that it is robust in general. The examples given in the end of the previous section suggest that classes of implementable surpluses can be very broad and bunching rather uncommon if the utility is concave in types. Some explicit examples solved below confirm this intuition.

The question of the robustness of bunching is not purely academic. Indeed, as first argued by Champsuar and Rochet (1989), if the outside option of the consumer is interpreted as her ability to purchase a substitute, bunching may result in the discontinuity of the market share of the monopolist in price and quality of the substitute. If now one tries to extend the model

---

[2] in the unidimensional case convexity of the type space is equivalent to its path connectedness, which is a topological property and therefore is preserved under diffeomorphisms. Hence, in the unidimensional case one can always select a diffeomorphism that will make $u(\cdot, \mathbf{x})$ convex without violating the convexity of Ω.

to the case of an oligopoly, the equilibrium may fail to exist. In practice this can be reflected in a high instability of the market. Therefore, it may be practically important to know the conditions under which bunching occurs and under which it does not. As I argued in this section, the latter possibility need not be atypical.

7.3 Path Independence of Information Rents

In the previous section I discussed the implementability of the surplus function and argued that it is a direct generalization of the monotonicity constraint for the unidimensional case. However, in multidimensional case implementability imposes additional constraints on the set of allocations, which are absent in the unidimensional case. The nature of these constraints is closely related to the notion of the *information rent.* Recall, that in the unidimensional case given an implementable allocation $\mathbf{x}(\cdot)$ we defined the information rent type α_2 earns over type α_1 in the unidimensional case by

$$I_{21} = \int_{\alpha_1}^{\alpha_2} u_{\alpha}(\alpha, \mathbf{x}(\alpha))d\alpha. \tag{7.45}$$

It would be natural to define the information rent in multidimensional case by

$$I_{21} = \int_{\Gamma} u_{\boldsymbol{\alpha}}(\boldsymbol{\alpha}, \mathbf{x}(\boldsymbol{\alpha}))d\boldsymbol{\alpha}, \tag{7.46}$$

where Γ is some smooth path connecting points α_1 and α_2. However, definition (7.46) immediately rises a question: For what allocations is the integral on the right hand side path independent? In economic terms the same question reads: For what allocations the information rent is well defined?

Recall from part one that the integral on the right hand side of (7.46) is path independent if and only if the vector field

$$b(\boldsymbol{\alpha}) = u_{\boldsymbol{\alpha}}(\boldsymbol{\alpha}, \mathbf{x}(\boldsymbol{\alpha})), \tag{7.47}$$

is conservative (I, Theorem 8). Now it is straightforward to notice the the envelope conditions for consumer optimization, which are equivalent to the local incentive compatibility, state that $b(\cdot)$ is a conservative vector field whose potential is equal to the consumer surplus. Therefore, the local incentive compatibility implies to the existence of well-defined (path independent) information rents.

7.4 Cost Based Tariffs

Before we start discussing different approaches to genuinely multidimensional screening problems it is useful to look at the class of problems that can be essentially reduced to a unidimensional model. Start with problem (7.2)-(7.3) and consider an auxiliary problem

$$\max u(\boldsymbol{\alpha}, \mathbf{x}) \tag{7.48}$$

$$s.t.\ c(\mathbf{x}) = y. \tag{7.49}$$

Namely we ask: given the monopolist is to spend amount y to serve type $\boldsymbol{\alpha}$ what is the best way to do it? Let $v(\boldsymbol{\alpha}, y)$ be the value of the problem. If it happens that $v(\alpha, y) = \phi(y)h(\boldsymbol{\alpha})$ and $f(\alpha) = f_1(\ h(\boldsymbol{\alpha}))$ the problem is reduced to the one-dimensional one, since all relevant information about the tastes collapses to the parameter $h(\boldsymbol{\alpha})$. (In fact, the condition is a little bit more general, it is allowed that $f(\alpha) = f_1(h(\boldsymbol{\alpha})) \times f_2(\boldsymbol{\alpha})$, where $f_2(\cdot)$ is homogenous of degree zero. See, Armstrong (1996) for the details). Note that the optimal tariff depends only on y, i. e., it is cost-based.

Example 183 *Let the individual's utility be given by:*

$$u(\alpha, x, t) = \alpha_1 x_1 + \alpha_2 x_2 + \sqrt{\alpha_1 \alpha_2} x_3 - t \tag{7.50}$$

and the cost of production is

$$c(x) = \frac{1}{2}(x_1^2 + x_2^2) + \frac{1}{4}x_3^2 \tag{7.51}$$

The set $\Omega = \{\alpha \in R_+^2 : \alpha_1 + \alpha_2 < b\}$. The distribution of types is given by the following density function:

$$f(\alpha_1, \alpha_2) = \frac{\exp(-\alpha_1 - \alpha_2)}{1 - (b+1)\exp(-b)}. \tag{7.52}$$

The value of the outside option is type independent and normalized to be zero. Define

$$v(\alpha, y) = \max_{x \in R_+^3} (\alpha_1 x_1 + \alpha_2 x_2 + \sqrt{\alpha_1 \alpha_2} x_3) \tag{7.53}$$

$$s.t. \frac{1}{2}(x_1^2 + x_2^2) + \frac{1}{4}x_3^2 = y. \tag{7.54}$$

The solution to this problem is

$$x_1 = \frac{\alpha_1 \sqrt{2y}}{\alpha_1 + \alpha_2} \tag{7.55}$$

$$x_2 = \frac{\alpha_2 \sqrt{2y}}{\alpha_1 + \alpha_2} \tag{7.56}$$

$$x_3 = x_1 = \frac{2\sqrt{\alpha_1 \alpha_2}\sqrt{2y}}{\alpha_1 + \alpha_2} \tag{7.57}$$

$$v(\alpha, y) = (\alpha_1 + \alpha_2)\sqrt{2y}. \tag{7.58}$$

where

$$\frac{1}{\lambda(\alpha, y)} = \frac{\sqrt{2y}}{(\alpha_1 + \alpha_2)}. \tag{7.59}$$

Therefore, the tariff is cost based. To find the tariff make a following change of variables

$$\begin{cases} \alpha_1 = \frac{1}{2}(\gamma + \xi) \\ \alpha_2 = \frac{1}{2}(\gamma - \xi) \end{cases}. \tag{7.60}$$

Then the type space is transformed into

$$\Omega' = \{(\gamma, \xi) \in (0, b) \times (-b, b) : -\gamma < \xi < \gamma\}$$

and, after carrying out integration with respect to ξ, *the monopolist's objective function becomes (up to a constant proportionality factor)*

$$\int_0^b (\gamma\sqrt{2y} - s - y)\gamma \exp(-\gamma) d\gamma. \tag{7.61}$$

Using any of the standard technique for the unidimensional case (see, Chapter 6) it can be shown that the optimal contract solves:

$$\max_y (\sqrt{2y}(\gamma - \frac{1 - H(\gamma)}{h(\gamma)}) - y), \tag{7.62}$$

where

$$h(\gamma) = \frac{\gamma \exp(-\gamma)}{1 - (b + 1)\exp(-b)} \tag{7.63}$$

and $H(\cdot)$ *is the corresponding cumulative distribution function. The solution is*

$$\sqrt{2y} = \gamma - \frac{\gamma + 1}{\gamma} + \frac{b + 1}{\gamma} \exp(\gamma - b). \tag{7.64}$$

One can check that $y(\cdot)$ *is increasing. Hence, the allocation is implementable. Returning to the original variables*

$$\begin{cases} x_1 = \alpha_1(1 - \frac{\alpha_1+\alpha_2+1}{(\alpha_1+\alpha_2)^2} + \frac{b+1}{(\alpha_1+\alpha_2)^2} \exp(\alpha_1 + \alpha_2 - b)) \\ x_2 = \alpha_2(1 - \frac{\alpha_1+\alpha_2+1}{(\alpha_1+\alpha_2)^2} + \frac{b+1}{(\alpha_1+\alpha_2)^2} \exp(\alpha_1 + \alpha_2 - b)) \\ x_3 = 2\sqrt{\alpha_1\alpha_2}(1 - \frac{\alpha_1+\alpha_2+1}{(\alpha_1+\alpha_2)^2} + \frac{b+1}{(\alpha_1+\alpha_2)^2} \exp(\alpha_1 + \alpha_2 - b)) \end{cases}. \tag{7.65}$$

The exclusion region is given by

$$\Omega_0 = \{\alpha \in R_+^2 : 1 - \frac{\alpha_1 + \alpha_2 + 1}{(\alpha_1 + \alpha_2)^2} + \frac{b + 1}{(\alpha_1 + \alpha_2)^2} \exp(\alpha_1 + \alpha_2 - b) \leq 0\}. \tag{7.66}$$

It is easy to see that the exclusion region is non-empty, since for small values of $\alpha_1 + \alpha_2$ *the first term can be neglected, and the second negative term dominates the third positive term, since* $e^b > b + 1$ *for any* $b \in R$. *It is also interesting to note that all types on the boundary* $\alpha_1 + \alpha_2 = b$ *are served efficiently. As* $b \to \infty$ *the solution becomes*

$$\begin{cases} x_1 = \alpha_1(1 - \frac{\alpha_1+\alpha_2+1}{(\alpha_1+\alpha_2)^2}) \\ x_2 = \alpha_2(1 - \frac{\alpha_1+\alpha_2+1}{(\alpha_1+\alpha_2)^2}) \\ x_3 = 2\sqrt{\alpha_1\alpha_2}(1 - \frac{\alpha_1+\alpha_2+1}{(\alpha_1+\alpha_2)^2}) \end{cases}. \tag{7.67}$$

The exclusion region in this case is

$$\Omega_0 = \{\alpha \in R_+^2 : \alpha_1 + \alpha_2 \leq \frac{1+\sqrt{5}}{2}\}. \tag{7.68}$$

7.5 Direct Approach and Its Limitations

Direct approach to the multidimensional screening problems was first developed by Armstrong (1996). Consider a relaxed multidimensional screening problem and assume that $\Omega = R_+^m$ (note that in this section Ω is not bounded), and $s_0(\boldsymbol{\alpha}) = 0$. Then $s(\mathbf{0}) = 0$. Indeed, according to the participation constraint $s(\mathbf{0}) \geq 0$ but envelope conditions (7.33) imply that for any $\boldsymbol{\alpha} \in R_+^m$ the surplus $s(\boldsymbol{\alpha})$ is defined by $s(\mathbf{0})$ and the allocation. The monopolist can affect $s(\mathbf{0})$ adding a constant to the tariff, which will not affect the incentive compatibility conditions, therefore she will adjust it to make $s(\mathbf{0})$ as small as possible.

Consider the following integral that appears in the monopolist's objective (7.34):

$$I = \int_{\Omega} s(\boldsymbol{\alpha}) f(\boldsymbol{\alpha}) \mathbf{d}\boldsymbol{\alpha}. \tag{7.69}$$

Define function $V(\cdot) : R_+ \to R$ by

$$V(t) = \int_{\Omega} s(t\boldsymbol{\alpha}) f(\boldsymbol{\alpha}) \mathbf{d}\boldsymbol{\alpha}, \tag{7.70}$$

then $V(0) = 0$, $V(1) = I$ and

$$V'(t) = \int_{\Omega} (\boldsymbol{\alpha}, \boldsymbol{\nabla} s(t\boldsymbol{\alpha})) f(\boldsymbol{\alpha}) \mathbf{d}\boldsymbol{\alpha}. \tag{7.71}$$

Now, using

$$V(1) - V(0) = \int_0^1 V'(t) dt, \tag{7.72}$$

one can write

$$I = \int_0^1 (\int_{\Omega} (\boldsymbol{\alpha}, \boldsymbol{\nabla} s(t\boldsymbol{\alpha})) f(\boldsymbol{\alpha}) \mathbf{d}\boldsymbol{\alpha}) dt. \tag{7.73}$$

From the envelope conditions (7.33)

$$I = \int_0^1 \Big(\int_\Omega (\boldsymbol{\alpha}, \boldsymbol{\nabla}_{\boldsymbol{\alpha}} u(t\boldsymbol{\alpha}, \mathbf{x}(t\boldsymbol{\alpha}))) f(\boldsymbol{\alpha}) \mathbf{d}\boldsymbol{\alpha} \Big) dt. \tag{7.74}$$

Let us make the following substitution

$$\boldsymbol{\beta} = t\boldsymbol{\alpha}. \tag{7.75}$$

Note that it takes Ω into itself. Under this transformation the term

$$\alpha_i \partial u / \partial \alpha_i (t\boldsymbol{\alpha}, \mathbf{x}(t\boldsymbol{\alpha})) \tag{7.76}$$

is transformed as

$$\alpha_i \frac{\partial u}{\partial \alpha_i}(t\boldsymbol{\alpha}, \mathbf{x}(t\boldsymbol{\alpha})) = \frac{1}{t}(t\alpha_i)\frac{\partial u}{\partial \alpha_i}(t\boldsymbol{\alpha}, \mathbf{x}(t\boldsymbol{\alpha})) = \frac{1}{t}\beta_i \frac{\partial u}{\partial \beta_i}(\boldsymbol{\beta}, \mathbf{x}(\boldsymbol{\beta})). \tag{7.77}$$

The integral with respect to $\boldsymbol{\alpha}$ will be transformed according to

$$\int_\Omega (\boldsymbol{\alpha}, \boldsymbol{\nabla}_{\boldsymbol{\alpha}} u(t\boldsymbol{\alpha}, \mathbf{x}(t\boldsymbol{\alpha}))) f(\boldsymbol{\alpha}) \mathbf{d}\boldsymbol{\alpha} = \int_\Omega t^{-m-1} (\boldsymbol{\beta}, \boldsymbol{\nabla}_{\boldsymbol{\beta}} u(\boldsymbol{\beta}, \mathbf{x}(\boldsymbol{\beta}))) f(\frac{\boldsymbol{\beta}}{t}) \mathbf{d}\boldsymbol{\beta}, \tag{7.78}$$

therefore

$$I = \int_\Omega (\boldsymbol{\beta}, \boldsymbol{\nabla}_{\boldsymbol{\beta}} u(\boldsymbol{\beta}, \mathbf{x}(\boldsymbol{\beta})) \Big(\int_0^1 t^{-m-1} f(\frac{\boldsymbol{\beta}}{t}) dt \Big) \mathbf{d}\boldsymbol{\beta}. \tag{7.79}$$

Finally, introducing $\tau = 1/t$ and defining

$$g(\beta) = \int_1^\infty \tau^{m-1} f(\boldsymbol{\beta}\tau) d\tau \tag{7.80}$$

one obtains

$$I = \int_\Omega (\boldsymbol{\beta}, \boldsymbol{\nabla}_{\boldsymbol{\beta}} u(\boldsymbol{\beta}, \mathbf{x}(\boldsymbol{\beta})) g(\boldsymbol{\beta}) \mathbf{d}\boldsymbol{\beta} \tag{7.81}$$

and, substituting it back into the monopolist's objective (7.34) we arrive at the following maximization problem:

$$\max_{x(\cdot)} \int_\Omega (u(\boldsymbol{\alpha}, \mathbf{x}) - c(\mathbf{x}) - (\boldsymbol{\alpha}, \boldsymbol{\nabla}_{\boldsymbol{\alpha}} u(\boldsymbol{\alpha}, \mathbf{x}(\boldsymbol{\alpha})) \frac{g(\boldsymbol{\alpha})}{f(\boldsymbol{\alpha}}) f(\boldsymbol{\alpha}) d\boldsymbol{\alpha}. \tag{7.82}$$

The procedure used to derive (7.82) is known as integration along ray. Unfortunately it does not take into account even the envelope constraints (7.33), which can be satisfied by the pointwise maximizer of (7.82) only by an accident. Therefore, we arrive at the following result:

Theorem 184 *The pointwise maximizer of (7.82), $\mathbf{x}(\boldsymbol{\alpha})$, solves the relaxed multidimensional screening problem if and only if the vector field $\mathbf{b}(\cdot)$ defined by*

$$b_i(\boldsymbol{\alpha}) = \frac{\partial u}{\partial \alpha_i}(\boldsymbol{\alpha}, \mathbf{x}(\boldsymbol{\alpha})) \tag{7.83}$$

is conservative.

The last condition is too strong and it is not satisfied usually for $m > 1$ as we will see shortly. For $m = 1$ expression (7.82) coincides with expression (6.21) obtained in the previous chapter for $s(0) = 0$, i. e. integration along rays is a generalization of integration by parts technique applied by Mussa and Rosen (1978) to the unidimensional case. To see it, assume $m = 1$ and calculate

$$g(\alpha) = \int_1^\infty f(\alpha\tau)d\tau = \frac{1 - F(\alpha)}{\alpha}. \tag{7.84}$$

Therefore, (7.82) can be rewritten as

$$\int_\Omega (u(\alpha, \mathbf{x}) - c(\mathbf{x}) - u_\alpha(\alpha, \mathbf{x}(\alpha)) \frac{1 - F(\alpha)}{f(\alpha})f(\alpha)d\alpha, \tag{7.85}$$

which is exactly expression (6.21).

Example 185 *Let the consumers' utility, cost, and distribution of types be given by (7.50), (7.51), and (7.52). The value of the outside option is type independent and normalized to be zero. Note that utility is homogeneous of degree one in types and assume that $b = \infty$, so $\Omega = R^2_+$. Then we can, following Armstrong's integration along rays approach, define a candidate solution as a solution to a pointwise maximization of the following expression*

$$(1 - \frac{g(\boldsymbol{\alpha})}{f(\boldsymbol{\alpha})})u(\boldsymbol{\alpha}, x) - c(x), \tag{7.86}$$

where

$$g(\boldsymbol{\alpha}) = \int_1^\infty tf(t\boldsymbol{\alpha})dt. \tag{7.87}$$

where

$$g(\boldsymbol{\alpha}) = \int_1^\infty tf(t\boldsymbol{\alpha})dt. \tag{7.88}$$

Evaluating $g(\cdot)$ as

$$g(\alpha) = \int_1^\infty t\exp(-t(\alpha_1 + \alpha_2))dt = \frac{\exp(-\alpha_1 - \alpha_2)}{(\alpha_1 + \alpha_2)^2}(\alpha_1 + \alpha_2 + 1). \tag{7.89}$$

one obtains

$$\frac{g(\boldsymbol{\alpha})}{f(\boldsymbol{\alpha})} = \frac{\alpha_1 + \alpha_2 + 1}{(\alpha_1 + \alpha_2)^2}. \tag{7.90}$$

Therefore,

$$x_1 = \alpha_1 (1 - \frac{\alpha_1 + \alpha_2 + 1}{(\alpha_1 + \alpha_2)^2}) \tag{7.91}$$

$$x_2 = \alpha_2 (1 - \frac{\alpha_1 + \alpha_2 + 1}{(\alpha_1 + \alpha_2)^2}) \tag{7.92}$$

$$x_3 = 2\sqrt{\alpha_1 \alpha_2} (1 - \frac{\alpha_1 + \alpha_2 + 1}{(\alpha_1 + \alpha_2)^2}). \tag{7.93}$$

A direct calculation of $\nabla_\alpha v(\alpha, x(\alpha))$ *shoes that*

$$\frac{\partial b_1(\boldsymbol{\alpha})}{\partial \alpha_2} = \frac{\partial u}{\partial \alpha_1}(\boldsymbol{\alpha}, x(\boldsymbol{\alpha})) = (\alpha_1 + \alpha_2)(1 - \frac{\alpha_1 + \alpha_2 + 1}{(\alpha_1 + \alpha_2)^2}) \tag{7.94}$$

$$\frac{\partial b_2(\boldsymbol{\alpha})}{\partial \alpha_1} = \frac{\partial u}{\partial \alpha_2}(\boldsymbol{\alpha}, x(\boldsymbol{\alpha})) = (\alpha_1 + \alpha_2)(1 - \frac{\alpha_1 + \alpha_2 + 1}{(\alpha_1 + \alpha_2)^2}), \tag{7.95}$$

therefore vector field $\mathbf{b}(\cdot)$ *is conservative and the found allocation solves the relaxed problem. Moreover, in this case we know that it also solves the complete problem.*

Note, however, that if we replace cost (7.51) by

$$c(x) = \frac{1}{2}(x_1^2 + x_2^2 + \beta x_3^2) \tag{7.96}$$

then for $\beta \neq 1/2$ the integration along the rays technique will produce a solution that is inconsistent with the envelope conditions. Moreover, the optimal tariff will no longer be cost based. I will return to this example later in this chapter.

7.6 Dual Approach for $m = n$

As we have seen in the previous section, the main obstacle for the application of the direct approach is its inability to take into account a consequence of local incentive compatibility constraints, namely the requirement that a certain vector field is conservative. A way to circumvent this difficulty is to exclude allocation $\mathbf{x}$ from the maximization problem altogether. For this purpose one can attempt to solve the envelope conditions (7.33) for the allocation in terms of the type and the gradient of the surplus. This is the main idea of the dual approach. An important limitation of this approach is that it has to assume $m = n$.

Dual approach to multidimensional screening problems was developed by Rochet and Chone (1998). They assumed that the consumer's preferences are given by

$$v(\alpha, x) = \sum_{i=1}^{m} \alpha_i x_i, \tag{7.97}$$

i. e. they are linear in types. This allowed them to arrive at the solution of the complete problem and made the analysis of the relaxed problem much easier. However, the analysis of the relaxed problem does not use linearity in any substantial way. I will retain this assumption for the expositional simplicity.

In the linear case equation (7.33) implies that

$$s_{\alpha_i} = x_i. \tag{7.98}$$

Recall also that in this case a surplus function is implementable if and only if it is continuous and convex and assume that the value of outside option $s_0(\cdot)$ is implementable, i. e. the outside option is a result of participation in an alternative economic mechanism. Then the monopolist solves:

$$\max \int_{\Omega} (\sum_{i=1}^{m} \alpha_i s_{\alpha_i} - s - c(\nabla s)) f(\boldsymbol{\alpha}) \mathbf{d}\boldsymbol{\alpha} \tag{7.99}$$

$$s.t.\ s(\boldsymbol{\alpha}) \ \geq \ s_0(\boldsymbol{\alpha}),\ s(\cdot) \text{ is convex.} \tag{7.100}$$

The type space Ω is assumed to be an open, bounded, convex set with a smooth boundary. Using the optimization by vector space methods technique presented in the part one of this book (I, Corollary 150) one arrives at the following lemma.

Lemma 186 *Let $s^*(\cdot) \in H^1(\Omega)$ solve (7.99)-(7.100). Then there exists a positive distribution $\eta(\cdot) \in H^1(\Omega)$ such that $s^*(\cdot)$ solves*

$$\max \int_{\Omega} [(\sum_{i=1}^{m} \alpha_i s_{\alpha_i} - s - c(\nabla s)) f + \eta(s - s_0)] \mathbf{d}\boldsymbol{\alpha} \tag{7.101}$$

$$s.t.\ s(\cdot) \text{ is convex.} \tag{7.102}$$

and the support of $\eta(\cdot)$ is contained in the set

$$E = \{\boldsymbol{\alpha} \in \Omega : s(\boldsymbol{\alpha}) = s_0(\boldsymbol{\alpha})\}. \tag{7.103}$$

Therefore, the monopolist's problem is reduced to a calculus of variations problem with a convexity constraint.

7.6.1 The Relaxed Problem

Let us drop we a moment the convexity constraint. Then the solution to the relaxed problem is given by the following theorem:

Theorem 187 *Let $s^*(\cdot) \in H^1(\Omega)$ solve (7.101), (7.102), where the convexity constraint is dropped. Then there exists a distribution η such that the following conditions hold:*

$$\begin{aligned} div(\boldsymbol{\alpha} - \nabla c(\nabla s))f(\boldsymbol{\alpha})) &= \eta(\boldsymbol{\alpha}) - f(\boldsymbol{\alpha}) && (7.104) \\ (\boldsymbol{\alpha} - \boldsymbol{\nabla} c(\boldsymbol{\nabla} s(\boldsymbol{\alpha}))) \cdot \mathbf{n}(\boldsymbol{\alpha}) &= 0 \text{ on } \partial\Omega. && (7.105) \\ \eta(\boldsymbol{\alpha}) &\geq 0,\ \eta(\boldsymbol{\alpha})(s(\boldsymbol{\alpha}) - s_0(\boldsymbol{\alpha})) = 0, && (7.106) \end{aligned}$$

One can easily recognize in this characterization the first order conditions for the calculus of variations problem with a free boundary. Integrating (7.104) over Ω, using the divergence theorem and boundary condition (7.105) one obtains

$$\int_\Omega \eta(\boldsymbol{\alpha}) \mathbf{d}\boldsymbol{\alpha} = \int_\Omega f(\boldsymbol{\alpha}) \mathbf{d}\boldsymbol{\alpha} = 1, \qquad (7.107)$$

therefore $\eta(\cdot)$ is a probability distribution. However, as we have already seen, $\eta(\cdot)$ cannot be in general assumed to be a function even in the unidimensional case. Therefore, notation $\eta(\alpha)$ is purely symbolic and has a literal meaning only in the points, where $\eta(\cdot)$ is regular (see I, Section 5.1 for a more detailed discussion).

7.6.2 An Alternative Approach to the Relaxed Problem

In the previous subsection we took into account the participation constraint by introducing a Lagrange multiplier $\eta(\cdot)$. An alternative approach is to demand that the function

$$F(t) = V(s + ht), \qquad (7.108)$$

achieves its maximum over $t \geq 0$ at $t = 0$, for any $h \geq 0$, such that $h \in H^1(\Omega)$. Here functional $V(\cdot)$ is defined by

$$V(s) = \int_\Omega (\sum_{i=1}^m \alpha_i s_{\alpha_i} - s - c(\nabla s))f(\boldsymbol{\alpha}). \qquad (7.109)$$

The first order condition is

$$F'(0)h \leq 0 \qquad (7.110)$$

for any $h \geq 0$, which is equivalent to the statement that the Gateaux directional derivative of $V(\cdot)$ in the direction of any increasing h is nonpositive. Therefore,

$$\begin{aligned} div(\boldsymbol{\alpha} - \nabla c(\nabla s))f(\boldsymbol{\alpha})) + f(\boldsymbol{\alpha}) &\geq 0, \text{ a.e. on } \Omega, && (7.111) \\ (\boldsymbol{\alpha} - \boldsymbol{\nabla} c(\boldsymbol{\nabla} s(\boldsymbol{\alpha}))) \cdot \mathbf{n}(\boldsymbol{\alpha}) &\leq 0, \text{ a.e. on } \partial\Omega. && (7.112) \end{aligned}$$

Note that at points, where $s(\boldsymbol{\alpha}) > s_0(\boldsymbol{\alpha})$ both conditions should holds as equalities because in that points one can drop the requirement that $h \geq 0$.

This formulation is the one used by Rochet and Chone (1998). Its advantage is that it deals with functions, rather then distributions. However, the cost of it is that the system of partial differential equations is now replaced with a system of partial differential inequalities. Which formulation one should use depends on the problem at hand. We will see below that the alternative formulation is very useful when characterizing the geometry of the participation region.

7.6.3 The Complete Problem

There are two approaches to the complete problem in the linear case. One uses the results on calculus of variations with convexity constraint obtained by Lions (1998), the other relies on the notion of the sweeping operator also known as the balayage. Both techniques were discussed in part one of this book. Let us start with giving the first order characterization using Lions results.

Theorem 188 *Let $\eta(\cdot)$ be a distribution that satisfies (7.106). Assume that $s^*(\cdot) \in H^1(\Omega)$ solves (7.101), (7.102). Then there exists a positively semidefinite matrix of distributions $\mu = \{\mu_{ij}\}$ such that*

$$div(\boldsymbol{\alpha} - \nabla c(\nabla s))f(\boldsymbol{\alpha})) = \eta(\boldsymbol{\alpha}) - f(\boldsymbol{\alpha}) + \sum_{i,j=1}^{m} \frac{\partial^2 \mu_{ij}}{\partial x_i \partial x_j}, \quad (7.113)$$

$$\int_{\Omega} \sum_{i,j=1}^{m} \mu_{ij} \frac{\partial^2 s_{ij}}{\partial x_i \partial x_j} \mathbf{dx} = 0, (\boldsymbol{\alpha} - \boldsymbol{\nabla} c(\boldsymbol{\nabla} s)) \cdot \mathbf{n} = 0 \ on\ \partial\Omega, (7.114)$$

where $\mathbf{n}$ is the unit vector normal to the boundary.

This Theorem follows immediately from (I, Theorem 151). I will use it to derive the characterization of the bunching region, provided by Rochet and Chone (1998). For that purpose assume $A \subset \Omega$ is a Borel set and define a charge $\nu(\cdot)$ by

$$\nu(A) = \int_A (div(\boldsymbol{\alpha} - \nabla c(\nabla s))f(\boldsymbol{\alpha}) - \eta(\boldsymbol{\alpha}) + f(\boldsymbol{\alpha}))\mathbf{d}\boldsymbol{\alpha}. \quad (7.115)$$

Let $h(\cdot)$ be arbitrary convex function. Multiply both sides of (7.113) on h and integrate over Ω to get

$$\nu h \equiv \int_{\Omega} h d\nu = \int_{\Omega} h \sum_{i,j=1}^{m} \frac{\partial^2 \mu_{ij}}{\partial x_i \partial x_j} = \int_{\Omega} \mu_{ij} \sum_{i,j=1}^{m} \frac{\partial^2 h}{\partial x_i \partial x_j} = \int_{\Omega} Tr(\mu D^2 h). \quad (7.116)$$

All derivatives here a taken in the sense of distributions and Tr is a trace of a matrix. Since both matrices μ and $D^2\ h$ are positively semi-definite the trace of their product is non-negative and therefore

$$\nu h \geq 0 \tag{7.117}$$

for any convex h. Let ν_+ and ν_- be two measures such that the charge ν can be uniquely represented in a form

$$\nu = \nu_+ - \nu_-, \tag{7.118}$$

then (7.117) implies that

$$\nu_+ h \geq \nu_- h \tag{7.119}$$

for any convex h and therefore by Cartier's Theorem (I, Theorem 143) there exists a balayage T such that

$$\nu_+ = T\nu_-. \tag{7.120}$$

Let Ω_B be the support of the distribution μ. Note that it coincides with the support of the charge ν. It can be partitioned into a union of bunches $\Omega(x)$ on each of which $s(\cdot)$ is linear. Moreover, the definition of the balayage and condition (7.120) imply that

$$\int\limits_{\Omega(x)} \boldsymbol{\alpha} d\nu(\boldsymbol{\alpha}) = \mathbf{0} \tag{7.121}$$

$$\int\limits_{\Omega(x)} d\nu(\boldsymbol{\alpha}) \quad = \quad 0. \tag{7.122}$$

These are the conditions derived by Rochet and Chone (1998). Below I derive the analogs of (7.121) and (7.122) in the nonlinear case. In the unidimensional case these conditions coincide with (6.54)-(6.56) and therefore fully characterize the solution to the complete problem. They, however, do not characterize the solution fully if $m \geq 2$ unless the problem has some additional symmetry.

7.6.4 The Geometry of the Participation Region

Let us assume that

$$s_0(\boldsymbol{\alpha}) = \boldsymbol{\alpha} \cdot \mathbf{x}_0 - t_0, \tag{7.123}$$

i. e. every consumer can purchase the good of quality $\mathbf{x}_0$ on the outside market at price t_0. Define $\boldsymbol{\alpha}_0$ by

$$\nabla c(\mathbf{x}_0) = \boldsymbol{\alpha}_0, \tag{7.124}$$

and assume that $\boldsymbol{\alpha}_0 \notin \overline{\Omega}$ and that the solution to the complete problem does not involve bunching. Let us also assume that the distribution of types is such that

$$(\boldsymbol{\alpha}_0 - \boldsymbol{\alpha}) \cdot \boldsymbol{\nabla} f(\boldsymbol{\alpha}) - (m-1) f(\boldsymbol{\alpha}) \leq 0 \tag{7.125}$$

for all $\alpha \in \Omega$. Then in the exclusion region two following conditions should hold:

$$div((\boldsymbol{\alpha} - \boldsymbol{\alpha}_0) f(\boldsymbol{\alpha})) + f(\boldsymbol{\alpha}) \geq 0 \tag{7.126}$$

$$(\boldsymbol{\alpha} - \boldsymbol{\alpha}_0) \cdot \mathbf{n}(\boldsymbol{\alpha}) \leq 0. \tag{7.127}$$

(Recall the alternative formulation of the relaxed problem is Subsection 7.6.2). The first of this conditions is equivalent to our assumption (7.125) and therefore holds automatically. To understand the second condition define the set

$$\partial_{-}\Omega = \{\boldsymbol{\alpha} \in \partial\Omega : (\boldsymbol{\alpha} - \boldsymbol{\alpha}_0) \cdot \mathbf{n}(\boldsymbol{\alpha}) \leq 0\}. \tag{7.128}$$

Geometrically, $\partial_{-}\Omega$ is the set of the boundary points that are *directly exposed* to $\boldsymbol{\alpha}_0$, i. e. the segment of the straight line connecting these points with $\boldsymbol{\alpha}_0$ does not contain any other points of set Ω. Equation (7.127) implies that if the boundary is smooth and strictly convex, the intersection of the boundary of Ω and the boundary of the participation region is exactly $\partial_{-}\Omega$.

7.6.5 A Sufficient Condition for Bunching

Suppose that the solution to the complete problem does not involve bunching. Then the results of the previous subsection imply that set Ω_{-} of types directly exposed to the outside option defined as a convex hull of $\partial_{-}\Omega$, is contained in the exclusion region. Therefore,

$$\int_{\Omega_{-}} div(\boldsymbol{\alpha} - \nabla c(\nabla s)) f(\boldsymbol{\alpha})) \mathbf{d}\boldsymbol{\alpha} + \int_{\partial\Omega_{-}} (\boldsymbol{\alpha} - \mathbf{\nabla c}(\mathbf{\nabla s}), \mathbf{n}) \mathbf{f}(\boldsymbol{\alpha}) \mathbf{d}\boldsymbol{\alpha}$$
$$\leq \int_{\Omega_0} \eta(\boldsymbol{\alpha}) \mathbf{d}\boldsymbol{\alpha} = 1, \tag{7.129}$$

Using equations (7.123) and (7.124) one obtains:

$$\int_{\Omega_{-}} div(\boldsymbol{\alpha} - \boldsymbol{\alpha}_0) f(\boldsymbol{\alpha})) \mathbf{d}\boldsymbol{\alpha} + \int_{\partial\Omega_{-}} (\boldsymbol{\alpha} - \boldsymbol{\alpha}_0, \mathbf{n}) \mathbf{f}(\boldsymbol{\alpha}) \mathbf{d}\boldsymbol{\alpha} \leq 1 \tag{7.130}$$

and therefore,

$$\int_{\Omega_{-}} div(\boldsymbol{\alpha} - \boldsymbol{\alpha}_0) f(\boldsymbol{\alpha})) \mathbf{d}\boldsymbol{\alpha} + \int_{\partial\Omega_{-}} (\boldsymbol{\alpha} - \boldsymbol{\alpha}_0, \mathbf{n}) \mathbf{f}(\boldsymbol{\alpha}) \mathbf{d}\boldsymbol{\alpha} > 1 \tag{7.131}$$

is sufficient for bunching.

The results of the last two subsections were first derived by Rochet and Chone (1998).

7.6.6 The Extension of the Dual Approach for $n > m$

Let

$$u(\boldsymbol{\alpha}, \mathbf{x}) = \sum_{i=1}^{m} \alpha_i u_i(\mathbf{x}), \tag{7.132}$$

where $\mathbf{x} \in R_+^n$ and $n > m$. As was first noted by Basov (2001), this problem can be reduced to one with $m = n$.

Let us start with interpreting the monopolist's problem in the following way: There are m different types of artificial goods, call them "utils." The ith util is produced from a vector of market goods $\mathbf{x}$, using a production function $u_i(\mathbf{x})$. Since the dimensionality of $\mathbf{x}$ is greater than the dimensionality of utils space, every possible combination of utils can be produced. Generically, there is a continuum of ways to produce the same combination of utils. Since, given the payment, both the incentive compatibility and the participation constraints depend only on the amount of utils obtained, and not on the vector $\mathbf{x}$ that implements them, the monopolist's problem can be divided into two parts. First, choose a vector of goods, $\mathbf{x}$, such that it minimizes the cost given the levels of util production. Second, solve for the optimal tariff as a function of utils.

To formulate the method described above more precisely, assume that the functions u_i are twice continuously differentiable, strictly quasiconcave, and satisfy

$$u_i(0) = 0 \tag{7.133}$$

$$\lim_{x_k \to \infty} u_i(\mathbf{x}) = \infty, \tag{7.134}$$

the Jacobi matrix, J, given by

$$J_{ij} = \frac{\partial u_i(\mathbf{x})}{\partial x_j} \tag{7.135}$$

has a full rank for all $\mathbf{x} \in R_+^n$, and the cost function $c(\cdot)$ is increasing, convex, twice continuously differentiable, and the eigenvalues of the matrix of its second derivatives are uniformly bounded from above and are uniformly bounded away from zero from below. Then the monopolist's problem can be solved in two steps. First, consider the problem:

$$\min c(\mathbf{x}) \tag{7.136}$$

$$s.t.\ u_i(\mathbf{x}) \geq u_i,\ i = 1, ..., n. \tag{7.137}$$

Given the assumptions, the solution to this problem exists and is unique. Denote this solution by $\mathbf{x}^*(\mathbf{u})$ and define $\theta(\mathbf{u}) = c(\mathbf{x}^*(\mathbf{u}))$. Then $\theta(\mathbf{u})$ will be convex, twice continuously differentiable, and the eigenvalues of the matrix of its second derivatives will be uniformly bounded from above and uniformly bounded away from zero from below. Now, the monopolist's problem can be rewritten in the form:

$$\begin{array}{l} \max_s \int [\sum_{i=1}^m \alpha_i u_i - \theta(\mathbf{u}) - s(\boldsymbol{\alpha})] f(\boldsymbol{\alpha}) d\alpha \\ s.t. s_{\alpha_i} = u_i, \ s(\cdot) - \text{convex}, \ s(\boldsymbol{\alpha}) \geq s_0(\boldsymbol{\alpha}) \end{array} \quad (7.138)$$

But this is exactly the problem studied in the last subsection. It is intuitively obvious that if $\mathbf{u}(\alpha)$ solves (7.138) then $\mathbf{x}^*(\mathbf{u}(\boldsymbol{\alpha}))$ solves the original problem. For the details of the proof, see Basov (2001).

7.7 Hamiltonian Approach and the First Order Conditions

In the previous section I described the dual approach to the multidimensional screening problem. Though it overcame some of the limitations of the direct approach, the assumptions under which it can be used are still rather restrictive. First, it requires that $m = n$. Even if this condition is satisfied it still may be hard to solve the envelope conditions explicitly for the allocation. In this section I am going to develop a method which will allow us to overcome these difficulties.

One has to consider two different cases. If $n \geq m$ then the problem the monopolist faces is a multidimensional optimal control problem with state variable $s(\cdot)$, vector of controls $\mathbf{x}$, and a system of independent evolution equations given by (7.33). Therefore, one can apply the Hamiltonian technique I describe below straight away. However, if $m > n$ the system (7.33) contains more equations that the control variables, therefore the equations in system (7.33) are dependent and one has to take this dependency directly into account.

The formal trick for doing this is the following. Introduce new variables z by formulae

$$z_i = \frac{\partial u}{\partial \alpha_i}(\alpha, \mathbf{x}), \ i = \overline{n+1, m} \quad (7.139)$$

$$z_i = x_i, \ i = \overline{1, n}. \quad (7.140)$$

Then z_i can be interpreted as new controls and the number of controls is now the same as the number of equations in the system (7.33). The controls, however, are dependent because of the equation (7.139).

One can give the following economic interpretation of this procedure. Interpret vector z as a vector of artificial goods, utils. If $m > n$ there are

m different types of utils, however, the set of feasible utils combination is an $n-$dimensional manifold in R_+^m. Hence, the problem with $m > n$ can be interpreted as a problem with $m = n$ subject to some production constraints. Geometrically it means that the set of feasible utils combinations is an $n-$dimensional manifold in R^m.

Define the relaxed problem by

$$\begin{gathered}\max_{\mathbf{z}\geq\mathbf{0}} \int_\Omega (u(\boldsymbol{\alpha},\mathbf{z}) - s(\boldsymbol{\alpha}) - c(\mathbf{z}))f(\boldsymbol{\alpha})\mathbf{d}\boldsymbol{\alpha} \\ s.t. \nabla s(\boldsymbol{\alpha}) = \nabla_{\boldsymbol{\alpha}} u(\boldsymbol{\alpha},\mathbf{z}),\ i=\overline{1,m}, \\ z_i = \partial u/\partial\alpha_i(\boldsymbol{\alpha},\mathbf{z}),\ i=\overline{n+1,m}, \\ s(\boldsymbol{\alpha}) \geq s_0(\boldsymbol{\alpha})\end{gathered} \tag{7.141}$$

Following the same steps we used while studying the dual approach we can arrive at the following result:

Lemma 189 *Let allocation $z^*(\cdot)$ solve (7.141) and $s^*(\cdot)$ be the corresponding surplus. Then there exists a distribution, $\eta(\cdot)$, such that $z^*(\cdot)$ solves*

$$\begin{gathered}\max \int_\Omega ((u(\boldsymbol{\alpha},\mathbf{z}) - s(\boldsymbol{\alpha}) - c(\mathbf{z}))f(\boldsymbol{\alpha}) + \boldsymbol{\eta}(\boldsymbol{\alpha})(\mathbf{s}(\boldsymbol{\alpha}) - \mathbf{s}_0(\boldsymbol{\alpha})))\mathbf{d}\boldsymbol{\alpha} \\ s.t. \nabla s(\boldsymbol{\alpha}) = \nabla_{\boldsymbol{\alpha}} u(\boldsymbol{\alpha},\mathbf{z}), i=\overline{1,m}, \\ z_i = \partial u/\partial\alpha_i(\boldsymbol{\alpha},\mathbf{z}),\ i=\overline{n+1,m},\end{gathered} \tag{7.142}$$

and $s^*(\cdot)$ is the corresponding surplus. Moreover,

$$\eta(\boldsymbol{\alpha}) \geq 0,\ \eta(\boldsymbol{\alpha})(s(\boldsymbol{\alpha}) - s_0(\boldsymbol{\alpha})). \tag{7.143}$$

Our next objective is to prove that $\eta(\cdot)$ is a probability distribution.

Lemma 190 *Let $s^*(\cdot)$ solve (7.142). Then*

$$\int_\Omega \eta(\boldsymbol{\alpha})\mathbf{d}\boldsymbol{\alpha} = 1. \tag{7.144}$$

Proof. Let

$$\int_\Omega \eta(\boldsymbol{\alpha})\mathbf{d}\boldsymbol{\alpha} = c. \tag{7.145}$$

Consider a pair $(s^*(\cdot)+\varepsilon,\ z^*(\cdot))$ for some constant ε. Then $(s^*(\cdot)+\varepsilon,\ z^*(\cdot))$ satisfy all the constraints of (7.142) but the value of the objective function changes by $\varepsilon(c-1)$. Therefore, if $c \neq 1$ one can increase the value of the of the objective by appropriately choosing ε. ■

Let $\mathbf{n}$ be the unit vector normal to $\partial\Omega$ and define the costate vector $\boldsymbol{\lambda}(\cdot)$ to be the solution to the following boundary problem:

$$div\boldsymbol{\lambda} = f(\boldsymbol{\alpha}) - \eta(\boldsymbol{\alpha}) \tag{7.146}$$

$$\boldsymbol{\lambda}\cdot\mathbf{n} = 0. \tag{7.147}$$

Recall from part one (I, Example 36) that such $\boldsymbol{\lambda}$ exists since

$$\int_{\Omega} (f(\boldsymbol{\alpha}) - \eta(\boldsymbol{\alpha}))\mathbf{d}\boldsymbol{\alpha} = 0. \tag{7.148}$$

Consider the following integral that appears in the monopolists objective function

$$\int_{\Omega} s(\boldsymbol{\alpha})(f(\boldsymbol{\alpha}) - \eta(\boldsymbol{\alpha}))\mathbf{d}\boldsymbol{\alpha} = \int_{\Omega} s(\boldsymbol{\alpha}) div \boldsymbol{\lambda} \mathbf{d}\boldsymbol{\alpha}. \tag{7.149}$$

Using the Gauss-Ostragrasky Theorem (I, Theorem 14), the identity

$$div(s\boldsymbol{\lambda}) = \boldsymbol{\lambda}\boldsymbol{\nabla} s + s div \boldsymbol{\lambda}, \tag{7.150}$$

equation (7.148), and the envelope conditions one obtains:

$$\int_{\Omega} s(\boldsymbol{\alpha})(f(\boldsymbol{\alpha}) - \eta(\boldsymbol{\alpha}))\mathbf{d}\boldsymbol{\alpha} = -\int_{\Omega} \boldsymbol{\lambda}\boldsymbol{\nabla}_{\boldsymbol{\alpha}} \mathbf{u}(\boldsymbol{\alpha}, \mathbf{z})\mathbf{d}\boldsymbol{\alpha}. \tag{7.151}$$

Finally, the monopolist's objective can be rewritten as

$$\int_{\Omega} ((u(\boldsymbol{\alpha}, \mathbf{z}) - c(\mathbf{z}))f(\boldsymbol{\alpha}) + \boldsymbol{\lambda}\boldsymbol{\nabla}_{\boldsymbol{\alpha}} \mathbf{u}(\boldsymbol{\alpha}, \mathbf{z}))\mathbf{d}\boldsymbol{\alpha}. \tag{7.152}$$

The monopolist will maximize the integrand pointwise subject to

$$z_i = \partial u / \partial \alpha_i(\boldsymbol{\alpha}, \mathbf{z}), i = \overline{n+1, m}. \tag{7.153}$$

Denoting the Lagrange multipliers on this constraints $\mu_j(\cdot)$ one arrives at the following first order characterization of the solution to the relaxed problem in the general case:

Theorem 191 *Suppose the surplus function $s^*(\cdot)$ solves the problem (7.141) Then there exist continuously differentiable vector function $\boldsymbol{\lambda} : \Omega \to R^m$, and distributions $\boldsymbol{\mu} : \Omega \to R^{m-n}$, $\eta : \Omega \to R_+$ such that and the following first order conditions hold:*

$$\begin{cases} div\boldsymbol{\lambda} = -\frac{\partial H}{\partial s} \ a.e. \ on \ \Omega. \\ \langle \boldsymbol{\lambda}, \mathbf{n} \rangle = 0 \ a.e. \ on \ \partial\Omega. \\ \eta \geq 0, \ s \geq s_0(\alpha) \\ \eta(s(\alpha) - s_0(\alpha)) = 0 \\ z \in \arg\max H(s, \mathbf{z}, \boldsymbol{\alpha}; \boldsymbol{\lambda}, \boldsymbol{\mu}) \end{cases} \tag{7.154}$$

where the Hamiltonian $H(s, \mathbf{z}, \boldsymbol{\alpha}; \boldsymbol{\lambda}, \boldsymbol{\mu})$ is defined by

$$H = (u - s - c)f + \boldsymbol{\lambda} \cdot \boldsymbol{\nabla}_{\alpha} u + \sum_{i=n+1}^{m} \mu_i (z_i - \frac{\partial u}{\partial \alpha_i}) + \eta(s - s_0). \tag{7.155}$$

7.7.1 The Economic Meaning of the Lagrange Multipliers

In a relaxed screening problem three different types of Lagrange multipliers appear: the Lagrange multiplier on the participation constraint, $\eta(\cdot)$, the vector of Lagrange multiplier on the local incentive compatibility constraints, $\boldsymbol{\lambda}(\cdot)$, and the vector $\boldsymbol{\mu}(\cdot)$, which represents the Lagrange multipliers on the utils production constraint. The economic meaning of $\eta(\cdot)$ and $\boldsymbol{\lambda}(\cdot)$ is obvious. The first is the marginal value for the monopolist from decreasing the outside option of type $\boldsymbol{\alpha}$, while the second is the marginal value from relaxing the local incentive compatibility constraint. Interpretation of $\boldsymbol{\mu}(\cdot)$ is, however, more interesting. Indeed, $\boldsymbol{\mu}(\cdot)$ should represent the marginal value of relaxing the utils production constraint. Recall, however, that the origin of this constraint is that the number of quality dimensions is smaller then the dimensionality of the consumer's type. Therefore, the only way to relax this constraint is to introduce new quality dimensions of the good. Hence, vector $\boldsymbol{\mu}(\cdot)$ measures the marginal value of the R&D in the product development.

7.8 Symmetry Analysis of the First Order Conditions

The system of the first order conditions is a nonlinear system of partial differential equations. There are no general ways to solve such a problem. Therefore, one has often to rely upon the numerical techniques. However, in some cases when a problem possesses additional symmetry one can use it to arrive at an explicit solution. In part one of this book I described the general techniques for a symmetry analysis of the systems of the partial differential equations. Here I will give an example of using this methods to solve a concrete screening problem.

Let the utility of a consumer be given by:

$$U = \alpha_1 x - \frac{1}{\gamma}(\alpha_2 + c)x^{\gamma} - t, \tag{7.156}$$

where α_1 and α_2 are distributed independently and uniformly on $(0, a) \times (0, b)$ (i. e. $a_1 = 0$, $a_2 = a$, $b_1 = c$, $b_2 = b + c$), $\gamma > 1$ and $c > b/2$. The cost of production is zero. In the case $\gamma = 2$, $a = b = c = 1$ this problem was first considered by Laffont, Maskin, and Rochet (1987) and revisited by Basov (2001). Case $\gamma = 3$ is considered in Basov (2004). The first order conditions for this problem are:

$$\left\{\begin{array}{c} s_2+\frac{1}{\gamma}s_1^\gamma=0 \\ \mu_1 s_1^{\gamma-1}+(\gamma-1)\mu s_1^{\gamma-2}s_{11}+\mu_2=3-\eta \\ \eta\geq 0,\ \eta s=0,\ \int\limits_\Omega \eta(\alpha)d\alpha=ab \\ \mu=c \text{ at } \alpha_2=0,\ \mu=b+c \text{ at } \alpha_2=b,\ \mu s_1^{\gamma-1}=a \text{ at } \alpha_1=a \end{array}\right. . \quad (7.157)$$

Let us start with calculate the symmetry group for this system. For this purpose, I will first calculate a symmetry group of the first equation of the system and then look for its subgroup that leaves the rest of the system invariant.

Let us consider pure coordinate transformations, i. e. transformations (2.167) with $\chi=0$. Then

$$\chi_{\{1\}}s_1^{\gamma-1}+\chi_{\{2\}}=0, \quad (7.158)$$

where $\chi_{\{1\}}$, $\chi_{\{2\}}$ are given by (2.169) and can be calculated to be:

$$\left\{\begin{array}{l} \chi_{\{1\}}=-s_1\theta_1^1-s_2\theta_1^2, \\ \chi_{\{2\}}=-s_1\theta_2^1-s_2\theta_2^2 \end{array}\right. . \quad (7.159)$$

Substituting (2.169) into (2.171) and taking into account equation

$$s_2+\frac{1}{\gamma}s_1^\gamma=0 \quad (7.160)$$

one obtains:

$$(\theta_1^1-\frac{1}{\gamma}\theta_2^2)s_1^\gamma-\frac{1}{\gamma}s_1^{2\gamma-1}\theta_1^2+s_1\theta_2^1=0. \quad (7.161)$$

Since (7.161) the only constraint on $s_1$is equation (7.160), which we have already taken into account, coefficients before different powers of s_1 should vanish simultaneously. Therefore,

$$\left\{\begin{array}{c} \theta_1^1-\frac{1}{\gamma}\theta_2^2=0 \\ \theta_1^2=0 \\ \theta_2^1=0 \end{array}\right. . \quad (7.162)$$

The last two equations of system (7.162) imply that θ^i depends only on α_i. Now, since θ_1^1 depends only on α_1, while θ_2^2 depends only on α_2, the first of the equation of system (7.162) implies that both of these derivatives are constant and finally the solution is given by:

$$\left\{\begin{array}{l} \theta^1=A(\alpha_1-\alpha_1^*) \\ \theta^2=A\gamma(\alpha_1-\alpha_2^*) \end{array}\right. , \quad (7.163)$$

where A, α_1^*, and α_2^* are arbitrary constants. The finite form of the symmetry group of equation (7.160) has the form

$$\begin{cases} \widetilde{\alpha_1} = \alpha_1^* + (\alpha_1 - \alpha_1^*)\exp(A\tau) \\ \widetilde{\alpha_2} = \alpha_2^* + (\alpha_2 - \alpha_2^*)\exp(A\gamma\tau) \\ \widetilde{s} = s \end{cases} . \tag{7.164}$$

Since we are interested in a symmetry group, put $\alpha_1^* = \alpha_2^* = 0$. It is straightforward to check that if one extends (7.164) by

$$\begin{cases} \widetilde{\alpha_1} = \beta\alpha_1,\ \widetilde{a} = \beta a \\ \widetilde{\alpha_2} = \beta^\gamma\alpha_2,\ \widetilde{b} = \beta^\gamma b,\ \widetilde{c} = \beta^\gamma c \\ \widetilde{\mu} = \beta^\gamma\mu, \end{cases} \tag{7.165}$$

where $\beta = \exp(A\tau)$ then system (7.157) remains invariant under this transformation. Therefore, the consumer surplus is an invariant of group (7.165) and solves:

$$\alpha_1\frac{\partial s}{\partial\alpha_1} + \gamma\alpha_2\frac{\partial s}{\partial\alpha_2} + a\frac{\partial s}{\partial a} + \gamma b\frac{\partial s}{\partial b} + \gamma c\frac{\partial s}{\partial c} = 0. \tag{7.166}$$

To find the general solution start of equation (7.166) write the system of the characteristics (see I, Subsection 2.1.3):

$$\frac{d\alpha_1}{\alpha_1} = \frac{d\alpha_2}{\gamma\alpha_2} = \frac{da}{a} = \frac{db}{\gamma b} = \frac{dc}{\gamma c} = \frac{ds}{0}. \tag{7.167}$$

Five independent first integrals of system (7.167) are

$$\begin{cases} b/c = C_1 \\ b/a^\gamma = C_2 \\ (\alpha_1 - \delta_1(b/c, b/a^\gamma))^\gamma/(\alpha_2 - \delta_2(b/c, b/a^\gamma)) = C_3 \\ (\alpha_2 - \delta_2(b/c, b/a^\gamma))/a^\gamma = C_4 \\ s = C_5 \end{cases} . \tag{7.168}$$

Therefore, the general solution of equation (7.166) has a form

$$s = s\left(\frac{(\alpha_1 - \delta_1(b/c, b/a^\gamma))^\gamma}{\alpha_2 - \delta_2(b/c, b/a^\gamma)}, \frac{\alpha_2 - \delta_2(b/c, b/a^\gamma)}{a^\gamma}, \frac{b}{c}, \frac{b}{a^\gamma}\right). \tag{7.169}$$

Let us introduce new variables

$$\begin{cases} \xi = (\alpha_1 - \delta_1 a)^\gamma/(\alpha_2 - \delta_2 b) \\ \zeta = (\alpha_2 - \delta_2 b)/a^\gamma \end{cases} , \tag{7.170}$$

where δ_1 and δ_2 are homogenous of zero degree functions of a, b, and c but do not depend on α_1 and α_2. Then omitting trivial parametric dependence one can write $s = s(\xi, \zeta)$. Equation (7.160) in these coordinates takes the form:

$$\xi^{\gamma-1}s_\xi((\gamma s_\xi)^{\gamma-1}-\xi^{2-\gamma})+\zeta s_\zeta=0. \tag{7.171}$$

Note that for s_ξ to remain finite as $\xi \to 0$ it should be that $s_\zeta = 0$. Therefore, the non-trivial solutions of (7.171) defined over entire Ω are given by

$$s(\xi,\zeta)=\frac{\gamma-1}{\gamma}\xi^{\frac{1}{\gamma-1}}+C, \tag{7.172}$$

where C is an arbitrary constant. Since surplus function (7.172) satisfies the envelope conditions and is convex, it is implementable. Using the envelope conditions one can calculate

$$x=(\frac{\alpha_1-\delta_1 a}{\alpha_2-\delta_2 b})^{\frac{1}{\gamma-1}}. \tag{7.173}$$

Note that allocation rule (7.173) still contains two free parameters δ_1 and δ_2. To find them substitute (7.172) and (7.173) into the second and the fourth equations of system (7.157) to obtain the following boundary valued problem for μ :

$$\begin{cases} (\alpha_1-\delta_1 a)\mu_1+(\alpha_2-\delta_2 b)\mu_2=3(\alpha_2-\delta_2 b)-\mu \\ \mu=c \text{ at } \alpha_2=0,\ \mu=b+c \text{ at } \alpha_2=b,\ \mu x^{\gamma-1}=a \text{ at } \alpha_1=a \end{cases} \tag{7.174}$$

First, note that a continuous μ could not satisfy these conditions. Indeed, consider point $(a,0)$. The boundary conditions imply that $x(a,\,0)=(a/c)^{1/(\gamma-1)}$. Therefore, there is no distortion at the bottom right point. If μ were continuous then $x^{\gamma-1}(a,b)=1/\mu=(a/(b+c))$, which is the efficient level. But the incentive compatibility constraint between types $(a,0)$ and (a,b) implies that it should but biased downwards.

The reason for the discontinuity of μ is non-smoothness of the boundary of set Ω. Since $x(\cdot,\cdot)$ is continuous inside the participation region, the solution to (7.174) should be sought separately in two regions separated by an isoquant passing through the point (a,b), the corner of Ω.

The general solution of the partial differential equation for μ has the form

$$\mu=\frac{3}{2}(\alpha_2-\delta_2 b)+\frac{\phi(\frac{\alpha_1-\delta_1 a}{\alpha_2-\delta_2 b})}{\alpha_2-\delta_2 b}, \tag{7.175}$$

where ϕ is arbitrary continuously differentiable function. At the neighborhood of point $(a,0)$ the following boundary conditions should be satisfied:

$$\mu=c \text{ at } \alpha_2=0,\ \mu x^{\gamma-1}=a \text{ at } \alpha_1=a. \tag{7.176}$$

The first boundary condition implies that $\phi=0$ and

$$\mu=c+\frac{3}{2}\alpha_2,\ \delta_2=-\frac{2c}{3b}.$$

Using the second boundary condition and (7.173) one obtains:

$$\delta_1 = \frac{1}{3}. \tag{7.177}$$

Therefore,

$$x = \left(\frac{\alpha_1 - \frac{a}{3}}{\alpha_2 + \frac{2c}{3}}\right)^{\frac{1}{\gamma-1}} \tag{7.178}$$

Note that $\mu(a, b) = c + \frac{3}{2}b \neq b + c$, therefore this solution cannot be extended to the region containing the upper boundary. The solution in that region is given by (7.174) subject to

$$\mu = c \text{ at } \alpha_2 = 0, \;\; \mu = b + c \text{ at } \alpha_2 = b. \tag{7.179}$$

It is straightforward to show that (7.174) and (7.179) imply

$$\left\{ \begin{array}{c} \delta_2 b = \frac{b-2c}{4} \\ \phi = -\frac{1}{32}(3b + 2c)(b - 2c) \end{array} \right. . \tag{7.180}$$

Therefore,

$$x = \left(\frac{4(\alpha_1 - \delta_1 a)}{4\alpha_2 - b + 2c}\right)^{\frac{1}{\gamma-1}}. \tag{7.181}$$

Using continuity of $x(\cdot, \cdot)$ at point (a, b) one obtains

$$\delta_1 = \frac{1}{2}. \tag{7.182}$$

and

$$\left\{ \begin{array}{c} x = 0 \text{ if } \alpha_1 \leq \frac{a}{2} \\ x = \left(\frac{4\alpha_1 - 2a}{4\alpha_2 - b + 2c}\right)^{\frac{1}{\gamma-1}} \text{ if } \alpha_1 \geq \frac{a}{2} \text{ and } (3b + 2c)\alpha_1 - 2a\alpha_2 \leq 2ac + ab \\ x = \left(\frac{3\alpha_1 - a}{3\alpha_2 + 2c}\right)^{\frac{1}{\gamma-1}} \text{ if } \alpha_1 \geq \frac{a}{2} \text{ and } (3b + 2c)\alpha_1 - 2a\alpha_2 \geq 2ac + ab \end{array} \right. . \tag{7.183}$$

The optimal tariff is determined by

$$t(x) = \max_{\alpha}(\alpha_1 x - \frac{1}{\gamma}(\alpha_2 + c)x^\gamma - s(\alpha)), \tag{7.184}$$

where $s(\cdot)$ is given by (7.185) with δ_1 and δ_2 given above and C determined from $t(0) = 0$. Therefore,

$$t(x) = \left\{ \begin{array}{c} \frac{a}{2}x - \frac{1}{\gamma}(\frac{c}{2} + \frac{b}{4})x^\gamma, \text{ if } x < x_* \\ \frac{a}{6}x_* - \frac{1}{\gamma}(\frac{c}{6} + \frac{b}{4})x_*^\gamma + \frac{a}{3}x - \frac{c}{3\gamma}x^\gamma, \text{ if } x \geq x_* \\ x_* = \left(\frac{2a}{3b+2c}\right)^{\frac{1}{\gamma-1}} \end{array} \right. . \tag{7.185}$$

If $\gamma = 2$ and $a = b = c = 1$ this solution coincides with one obtained by Laffont, Maskin, and Rochet (1987). Note that as $\gamma \to \infty$ the tariff becomes piecewise affine

$$t(x) = \left\{ \begin{array}{c} \frac{a}{2}x \text{ if } x < 1 \\ \frac{a}{6} + \frac{a}{3}x \text{ if } x \geq 1. \end{array} \right. . \tag{7.186}$$

The exclusion region has a form

$$\Omega_0 = \{\alpha_1 \in (0,1) : \alpha_1 < \frac{a}{2}\}$$

and does not depend on b, c, and γ. This can be seen immediately from the form of the utility, since for any differentiable at zero tariff $x > 0$ if and only if $\alpha_1 > t'(0)$. Using this fact, system (7.157) and finding μ from (7.175) and (7.180) one can arrive at the exclusion region in a different way. Indeed, (7.157), (7.175), (7.180) together with the condition $x = 0$ imply

$$\eta(\alpha) = \frac{3}{2} + \frac{1}{2}\frac{(3b + 2c)(b - 2c)}{(4\alpha_2 + 2c - b)^2}.$$

Using the normalization condition

$$\int_0^{\alpha_1^*} d\alpha_1 (\int_0^b [\frac{3}{2} + \frac{1}{2}\frac{(3b + 2c)(b - 2c)}{(4\alpha_2 + 2c - b)^2}] d\alpha_2) = ab$$

one can calculate α_1^* to be equal to $a/2$.

7.9 The Hamiltonian Approach to the Complete Problem

Though the solution of the complete problem in the general case is still out of reach, one may arrive at conditions that hold on the bunch that generalize conditions (6.54)-(6.56) for $m = 1$ and conditions (7.121)-(7.122) obtained by Rochet and Chone (1998) in multidimensional case. Let us for simplicity assume that

$$s_0(\alpha) = v(\alpha, x_0) - t_0,$$

i. e. the consumers have accesses to an outside good of quality x_0 at price t_0. This assumption will not affect the results significantly. If the solution to the relaxed problem is not implementable the optimal allocation will include bunches. The definition of the bunch is, however, tricky if $m > n$. Indeed, if this is the case, for any allocation $x(\alpha)$ and almost all $\overline{x} \in X$ there will exist an $m-n$ dimensional manifold $M(\overline{x})$ embedded in Ω such that all consumers with types in $M(\overline{x})$ will purchase the same quality $\overline{x}$. I will call a bunch *genuine* if its dimension is at least $m-n+1$. I will call an allocation without genuine bunches a *separating* allocation. For any allocation $x(\alpha)$ the types space Ω can be partitioned in three disjoint regions as

$$\Omega = \Omega_0 \cup \Omega_B \cup \Omega_1$$

such that $x(\alpha) = x_0$ if $\alpha \in \Omega_0$, type α belongs to a genuine bunch if $\alpha \in \Omega_B$, and the allocation is separating for types in Ω_1. The bunching region Ω_B can itself be partitioned into bunches

$$\Omega_B = \underset{\overline{x}}{\cup} \Omega(\overline{x}),$$

where $\Omega(\overline{x})$ is a set of types of a dimension at least $m - n + 1$ such that all consumers with these types get the same allocation $\overline{x}$. In region Ω_1 the solution of the complete problem is described by system (7.154). To find some conditions that hold on a bunch, note that a bunch $\Omega(\overline{x})$ is characterized by the condition

$$z(\alpha) = \overline{x}, \ i = \overline{1, n.}$$

Adding this constraint to the optimization problem and following the same logic as in Theorem 192, one can write the monopolist's objective as

$$\max_{z, \overline{x}} \int_{\Omega} \{[v(\alpha, z) - c(z)]f(\alpha) + \langle \lambda, \nabla_\alpha v \rangle + \sum_{k=n+1}^{m} \mu_k (z_k - \frac{\partial v}{\partial \alpha_k})\} d\alpha$$
$$+ \int_{\Omega(\overline{x})} (z(\alpha) - \overline{x}) d\kappa(\alpha), \tag{7.187}$$

where $\kappa(\alpha)$ is a distribution supported by the bunch. The first order conditions with respect to $z(\alpha)$ are

$$\int_{\Omega} \frac{\partial}{\partial z_i} \{[v(\alpha, z) - c(z)]f(\alpha) + \langle \lambda, \nabla_\alpha v \rangle + \sum_{k=n+1}^{m} \mu_k (z_k - \frac{\partial v}{\partial \alpha_k})\} d\alpha = \int_{\Omega(\overline{x})} d\kappa(\alpha), \tag{7.188}$$

while the first order conditions with respect to $\overline{x}$ are

$$\int_{\Omega(\overline{x})} d\kappa(\alpha) = 0. \tag{7.189}$$

Conditions (7.188)-(7.189) imply that

$$\int_{\Omega} \frac{\partial}{\partial z_i} \{[v(\alpha, z) - c(z)]f(\alpha) + \langle \lambda, \nabla_\alpha v \rangle + \sum_{k=n+1}^{m} \mu_k (z_k - \frac{\partial v}{\partial \alpha_k})\} d\alpha = 0. \tag{7.190}$$

Since the integrand in (7.190) is identically zero outside the bunching region this condition can be rewritten as

$$\int_{\Omega_B} \frac{\partial}{\partial z_i} \{[v(\alpha, z) - c(z)]f(\alpha) + \langle \lambda, \nabla_\alpha v \rangle + \sum_{k=n+1}^{m} \mu_k (z_k - \frac{\partial v}{\partial \alpha_k})\} d\alpha = 0. \tag{7.191}$$

Since the monopolist can vary $\overline{x}$ on each bunch independently

$$\int_{\Omega(\overline{x})} \frac{\partial}{\partial z_i}\{[v-c]f(\alpha) + \langle \lambda, \nabla_\alpha v\rangle + \sum_{k=n+1}^{m} \mu_k(z_k - \frac{\partial v}{\partial \alpha_k})\}d\alpha = 0. \quad (7.192)$$

Condition (7.192) together with the smooth pasting condition that states that $x(\cdot)$ is continuous on the boundary of each bunch generalize (7.121)-(7.122) and (6.54)-(6.56).

7.10 Examples and Economic Applications

The purpose of the first example of this Section is to illustrate that when the GSC property fails the optimal allocation can be implemented by several different tariffs though the allocation itself remains unique.

Example 192 *Let $m = 1$, $n = 3$. Assume that the consumers' utilities are given by*

$$u(\alpha, \mathbf{x}) = \alpha(x_1 + x_2 + x_3), \quad (7.193)$$

while the cost of the monopolist is given by:

$$c(\mathbf{x}) = \frac{1}{2}(x_1^2 + x_2^2 + x_3^2). \quad (7.194)$$

That is all quality dimensions are perfect substitutes and the consumer only differ in their marginal rate of substitution between these goods and money. Assume that $\Omega = (0,1)$ and the types are distributed uniformly on Ω. Note that the GSC does not hold, since

$$\frac{\partial u}{\partial \alpha}(\alpha, \mathbf{x}) = \frac{\partial u}{\partial \alpha}(\alpha, \mathbf{y}) \Leftrightarrow x_1 + x_2 + x_3 = y_1 + y_2 + y_3 \nRightarrow \mathbf{x} = \mathbf{y}.$$

To find the optimal allocation, first solve

$$\min c(\mathbf{x})$$

$$s.t.\ x_1 + x_2 + x_3 = y.$$

The result is

$$x_1 = x_2 = x_3 = \frac{y}{3}.$$

$$c(y) = \frac{1}{6}y^2.$$

Now, the monopolist solves

$$\max \int_0^1 (\alpha y - s - \frac{1}{6}y^2)d\alpha = \int_0^1 (\alpha y - (1-\alpha)y - \frac{1}{6}y^2)d\alpha.$$

The solution is

$$y(\alpha) = 6\alpha - 3$$

and

$$x_1(\alpha) = x_2(\alpha) = x_3(\alpha) = 2\alpha - 1.$$

Types with $\alpha < 1/2$ *are excluded from the contract. Note that this allocation can be implemented by several tariffs. For example, a cost based tariff*

$$t(x) = \frac{1}{2}((\sqrt{\frac{x_1^2 + x_2^2 + x_3^2}{2}} + 1)^2 - 1)$$

implements this allocation. It can also be implemented by a tariff

$$\tau(x) = \frac{1}{2}(\frac{(x_1 + x_2 + x_3 + 3)^2}{6} - 1).$$

The last tariff implements the above allocation only weakly, since any triple (x_1, x_2, x_3) *satisfying*

$$x_1 + x_2 + x_3 = 6\alpha - 3$$

is an optimal choice for a consumer of type α *given the tariff.*

The next example shows how one can solve a multidimensional model with bunches. It was first considered by Wilson (1993) and revisited by Rochet and Chone (1998). I revisit this example using the techniques developed in this book.

Example 193 *Let the individual's utility be given by:*

$$u(\alpha, x, t) = \alpha_1 x_1 + \alpha_2 x_2 - t \tag{7.195}$$

and the cost of production is

$$c(x_1, x_2) = \frac{1}{2}(x_1^2 + x_2^2). \tag{7.196}$$

Therefore, the goods are perfect substitutes as in previous example, but now not only the marginal rate of substitution between each good and money but also the marginal rate of substitution between goods is private information. The type is distributed on $\Omega = (a, +\infty) \times (a, +\infty)$ *with a density*

$$f(\alpha_1, \alpha_2) = \exp(2a - \alpha_1 - \alpha_2) \tag{7.197}$$

for some $a > 1$*. The value of the outside option is independent of type and normalized to be zero. The envelope conditions are*

$$s_{\alpha_1} = x_1 \tag{7.198}$$

$$s_{\alpha_2} = x_2 \tag{7.199}$$

and the first order conditions are

$$\lambda_1 = (x_1 - \alpha_1)e^{2a-\alpha_1-\alpha_2} \quad (7.200)$$
$$\lambda_2 = (x_2 - \alpha_2)e^{2a-\alpha_1-\alpha_2} \quad (7.201)$$
$$div\boldsymbol{\lambda} = 3 - \eta e^{\alpha_1+\alpha_2-2a} + Tr(D^2\mu) \quad (7.202)$$
$$\lambda_i = 0 \text{ for } \alpha_i = a \quad (7.203)$$
$$\lim_{\alpha_i\to\infty} \lambda_i e^{2a-\alpha_1-\alpha_2} = 0 \quad (7.204)$$
$$\int_\Omega \eta(\boldsymbol{\alpha})\mathbf{d}\boldsymbol{\alpha} = 1, \eta \geq 0, \eta s = 0. \quad (7.205)$$

The distributions μ_{ij} *are supported by the bunching region. In the exclusion region* $x_1 = x_2 = 0$. *Therefore,*

$$\lambda_i = -\alpha_i \exp(2a - \alpha_1 - \alpha_2)\theta(\alpha_i - a), \quad (7.206)$$

where $\theta(\cdot)$ *is the Heviside step function, which takes into account the boundary conditions on the lower boundary of* Ω. *Therefore,*

$$\eta(\boldsymbol{\alpha}) = \exp(2a - \alpha)(3 - \alpha + a\delta(\alpha_1 - a) + a\delta(\alpha_2 - a)), \quad (7.207)$$

where $\alpha = \alpha_1 + \alpha_2$. *Since the problem is symmetric is* α_1 *and* α_2 *and the exclusion region is convex let us look for the exclusion region in the form*

$$\Omega_0 = \{\boldsymbol{\alpha} \in \Omega : \alpha \leq \alpha_0\}, \quad (7.208)$$

where α_0 *can be found from*

$$\int_\Omega \eta(\boldsymbol{\alpha})\mathbf{d}\boldsymbol{\alpha} = 1, \eta s = 0 \quad (7.209)$$

to be given by:

$$\alpha_0 = \frac{1}{2} + a + \sqrt{(a - \frac{1}{2})^2 + 1}. \quad (7.210)$$

In the full separation region the surplus solves

$$\Delta s - \frac{\partial s}{\partial \alpha_1} - \frac{\partial s}{\partial \alpha_1} = 3 - \alpha_1 - \alpha_2 \quad (7.211)$$

$$\frac{\partial s}{\partial \alpha_1} = a \text{ for } \alpha_i = a \quad (7.212)$$
$$\lim_{\alpha_i\to\infty} (\frac{\partial s}{\partial \alpha_i} - a) \exp(2a - \alpha_1 - \alpha_2) = 0. \quad (7.213)$$

Suppose that the separation region has a form

$$\Omega_1 = \{\boldsymbol{\alpha} \in \Omega : \alpha \geq \alpha^*\}. \quad (7.214)$$

Since the line $\alpha = \alpha^*$ *is the boundary of the bunching region,* x_i *has the same value along it. Since at points where the line crosses the axes* $x_i = a$ *it should have the same value at all points. Now, using the condition for existence of a solution for a Neumann boundary problem:*

$$\int_{\Omega_1} f(\boldsymbol{\alpha})d\boldsymbol{\alpha} = \int_{\partial\Omega_1} (\boldsymbol{\alpha} - \nabla s, \mathbf{n}) f(\boldsymbol{\alpha}) d\boldsymbol{\alpha} \tag{7.215}$$

one can find

$$\alpha^* = 2a + \frac{1+\sqrt{5}}{2}. \tag{7.216}$$

It is straightforward to check that $\alpha^* > \alpha_0$ *for any* $a > 0$*. Finally, in bunching region* $\Omega_B = \cup_x \Omega(x)$ *the solution has a form* $x_i = x_i(\alpha)$ *and the envelope conditions imply that* $s = s(\alpha_1 + \alpha_2)$*. Hence,*

$$x_1(\alpha) = x_2(\alpha) = s'(\alpha) \tag{7.217}$$

equations (7.202) and (7.192) imply that

$$3 - \alpha + 2x'(\alpha) + 2x(\alpha) + x - a = 0. \tag{7.218}$$

This, together with the smooth pasting conditions imply

$$x = \frac{1}{2}(\alpha - 1 - \frac{1}{\alpha - 2a}). \tag{7.219}$$

Solution in the region Ω_1 *should be found numerically.*

In the previous example the utility was linear in types and the optimal solution included a bunching region in accordance with the intuition provided by Rochet and Chone (1998). In the following example the utility is not convex in types and there is no bunching for some open set of the parameter values.

Example 194 *Let the individual's utility be given by:*

$$u(\alpha, x, t) = \alpha_1 x_1 + \alpha_2 x_2 + \sqrt{\alpha_1 \alpha_2} x_3 - t \tag{7.220}$$

and the cost of production is

$$c(x) = \frac{1}{2}(x_1^2 + x_2^2 + \beta x_3^2). \tag{7.221}$$

The set $\Omega = \{\alpha \in R_+^2 : \alpha_1 + \alpha_2 < b\}$*. Note that since*

$$\frac{\partial^2 u}{\partial \alpha_1^2} = -\frac{1}{2}\alpha_1^{-3/2}\alpha_2^{1/2} x_3 < 0 \tag{7.222}$$

the utility is not convex in types. The distribution of types is given by

$$f(\alpha_1, \alpha_2) = \frac{\exp(-\alpha_1 - \alpha_2)}{1 - (b+1)\exp(-b)}. \tag{7.223}$$

The value of the outside option is type independent and normalized to be zero.

We know from Example 184 that for $\beta = 1/2$ the tariff is cost based. Let us consider the general case. Start with considering the following problem:

$$\max(\alpha_1 x_1 + \alpha_2 x_2 + \sqrt{\alpha_1 \alpha_2} x_3) \tag{7.224}$$
$$s.t.c(x) = y. \tag{7.225}$$

The solution to this problem is given by:

$$x_1 = \frac{\alpha_1 \sqrt{2y}}{\alpha_1 + \alpha_2} \tag{7.226}$$
$$x_2 = \frac{\alpha_2 \sqrt{2y}}{\alpha_1 + \alpha_2} \tag{7.227}$$
$$x_3 = x_1 = \frac{2\sqrt{\alpha_1 \alpha_2}\sqrt{2y}}{\alpha_1 + \alpha_2} \tag{7.228}$$
$$v(\alpha, y) = (\alpha_1 + \alpha_2)\sqrt{2y}. \tag{7.229}$$

where

$$\frac{1}{\lambda(\alpha, y)} = \frac{\sqrt{2y}}{\sqrt{\alpha_1^2 + \alpha_2^2 + \frac{\alpha_1 \alpha_2}{\beta}}}. \tag{7.230}$$

Following Armstrong (1996), we conclude that the optimal tariff is cost based if and only if $u(\alpha, y) = v(y)\varphi(\alpha_1 + \alpha_2)$ which happens for $\beta = 1/2$. Therefore, $\beta = 1/2$ is the only case when the tariff is cost based and the solution in that case is given by (7.65). For an arbitrary value of β one has to use the Hamiltonian approach developed by Basov (2002) and described in the previous section. The envelope conditions for the consumer surplus are

$$\frac{\partial s}{\partial \alpha_1} = x_1 + \frac{1}{2}\sqrt{\frac{\alpha_2}{\alpha_1}} x_3 \tag{7.231}$$
$$\frac{\partial s}{\partial \alpha_2} = x_2 + \frac{1}{2}\sqrt{\frac{\alpha_1}{\alpha_2}} x_3 \tag{7.232}$$

The first order conditions have the form

$$\lambda_1 = (x_1 - \alpha_1)\exp(-\alpha_1 - \alpha_2) \tag{7.233}$$

$$\lambda_2 = (x_2 - \alpha_2)\exp(-\alpha_1 - \alpha_2) \tag{7.234}$$

$$x_3 = \frac{\exp(\alpha_1 + \alpha_2)}{\beta}(\sqrt{\alpha_1\alpha_2} + \frac{1}{2}(\lambda_1\sqrt{\frac{\alpha_2}{\alpha_1}} + \lambda_2\sqrt{\frac{\alpha_1}{\alpha_2}}) \tag{7.235}$$

$$div\lambda = \exp(-\alpha_1 - \alpha_2) - \eta \tag{7.236}$$

$$\lambda_1 + \lambda_2 = 0 \; at \; \alpha_1 + \alpha_2 = b \tag{7.237}$$

$$\lambda_1 = 0 \; at \; \alpha_1 = 0 \tag{7.238}$$

$$\lambda_2 = 0 \; at \; \alpha_2 = 0. \tag{7.239}$$

Within the participation region $\eta = 0$, let us look for a candidate solution of the form:

$$\lambda_i = \alpha_i \exp(-\alpha_1 - \alpha_2)\varphi(\alpha_1 + \alpha_2), \tag{7.240}$$

where φ is some continuously differentiable function. If we denote $z = \alpha_1 + \alpha_2$ then

$$div\lambda = \exp(-\alpha_1 - \alpha_2) \Rightarrow z\varphi' + (2 - z)\varphi = 1.$$

The condition on the boundary $\alpha_1 + \alpha_2 = b$ implies $\varphi(b) = 0$, while the two other boundary conditions are always satisfied. It is straightforward to check that the solution is given by

$$\varphi(z) = -\frac{1}{z} - \frac{1}{z^2} + (b+1)\frac{\exp(z-b)}{z^2}. \tag{7.241}$$

Hence

$$\lambda_i = -\alpha_i \exp(-\alpha_1 - \alpha_2)(\frac{\alpha_1 + \alpha_2 + 1}{(\alpha_1 + \alpha_2)^2} - \frac{b+1}{(\alpha_1 + \alpha_2)^2}\exp(\alpha_1 + \alpha_2 - b)). \tag{7.242}$$

and

$$\begin{cases} x_1 = \alpha_1(1 - \frac{\alpha_1+\alpha_2+1}{(\alpha_1+\alpha_2)^2} + \frac{b+1}{(\alpha_1+\alpha_2)^2}\exp(\alpha_1 + \alpha_2 - b)) \\ x_2 = \alpha_2(1 - \frac{\alpha_1+\alpha_2+1}{(\alpha_1+\alpha_2)^2} + \frac{b+1}{(\alpha_1+\alpha_2)^2}\exp(\alpha_1 + \alpha_2 - b)) \\ \frac{\sqrt{\alpha_1\alpha_2}}{\beta}(1 - \frac{\alpha_1+\alpha_2+1}{(\alpha_1+\alpha_2)^2} + \frac{b+1}{(\alpha_1+\alpha_2)^2}\exp(\alpha_1 + \alpha_2 - b)) \end{cases}. \tag{7.243}$$

But this is exactly allocation (7.65) and the integrability constraints imply that this allocation is implementable if and only if $\beta = 1/2$. Otherwise, $s(\cdot)$ solves a boundary value problem for an elliptic PDE:

$$\begin{cases} \sum_{i=1}^{2}(\frac{\partial(x_i(\alpha,\nabla s)-\alpha_i)}{\partial\alpha_i} - x_i(\alpha,\nabla s) + \alpha_i) = 1 \\ \sum_{i=1}^{2} x_i(\alpha,\nabla s) = b \; for \; \alpha_1 + \alpha_2 = b \end{cases}. \tag{7.244}$$

Therefore, we see that the Hamiltonian approach can be applied for arbitrary values of parameters. In general, the boundary value problem should be solved numerically. For β sufficiently close to $1/2$ its solution is implementable.

Note that the allocation in the last example is fully separating . Moreover, one can argue that there exists an open set of parameter values (β sufficiently close to $1/2$) for which the solution to this problem entails no bunching. This illustrates the general point that bunching need not be prevalent in the multidimensional case and arises from interaction of multidimensionality and joint convexity.

7.11 Exercises

1. Prove that for $m = 1$ conditions (7.121)-(7.122) imply (6.54)-(6.56).
2. Prove that (7.192) supplemented with the smooth pasting condition reduces to (7.121)-(7.122) if $m = n$ and

$$u(\alpha, x) = \sum_{i=1}^{n} \alpha_i x_i \tag{7.245}$$

and to (6.54)-(6.56) for $m = 1$.

7.12 Bibliographic Notes

The material of chapter still did not find its way into textbooks or monographs. Therefore, the main results can be found only in journal articles and articles in the collective volumes.

The first example of an explicitly solved two-dimensional screening model is Maskin, Laffont, and Rochet (1987). The example is Section 8 of this chapter is a direct generalization of that example. The multidimensional screening problem was first formulated in a general form by Wilson (1993) and Armstrong (1996). In the latter paper the first general result in the area, the genericity of exclusion, was derived. Section 1 of this chapter is devoted to this result. Armstrong (1996) also developed the integration across rays technique and characterized the class of problems for which the tariff is cost based. These results are presented in Sections 4 and 5.

Sections 2 and 3 of this Chapter describe implementability conditions in multidimensional models. These results in full generality were first obtained by Carlier (2002) and Basov (2002). For the case of the utilities linear in types the result was first obtained by Rochet (1987). It is also worth mentioning the early paper of McAfee and McMillan (1988) on the multidimensional incentive compatibility and the paper of Jehiel, Moldovanu, and Stacchetti (1999) on implementability in auctions with externalities. The result in the latter paper, though more general in the sense that it allows for externalities, allows only for the utilities that are linear in types and quasilinear in money. Therefore, it is a direct generalization of Rochet (1987) implementability condition in a direction different from one taken in the current work.

Section 6 discusses the dual approach to screening problem, developed by Rochet and Chone (1998). Sections 7 to 9 discuss the Hamiltonian approach. For the description of the Hamiltonian approach in the linear case, see Basov (2001). Development of the Hamiltonian approach in the nonlinear case is contained in Basov (2002, 2004).

A good review of the developements in the area by the turning of the century is Rochet and Stole (2003).

8

Beyond the Quasilinear Case

So far, we have concentrated on the case, when the utility of a consumer is quasilinear in money. This case received the most attention in the literature. However, one can easily give some economically interesting examples, where the most natural formulation leads to a consumer's utility, which is not quasilinear in money. Consider, for example, the following model of grant allocation (Bardsley and Basov, 2004). Risk averse institutions compete for grants for completing a research project. A project, if successful, will result in the provision of a public good whose value to the society is equal to one. Different institutions have projects that differ in the cost of completion and the probability of success. The government can choose an up-front payment and the prize in the case of success and is interested in maximizing the benefits of the society minus the completion costs. The institutions are assumed to be politically small, so there expected profits do not enter the government's objective. If one denotes the cost of the project c, the probability of success q, the up-front payment t and the prize for success x, the utility of the institution conditional on participation in the government's scheme will be

$$u(c, a; x, t) = qu(t + x - c) + (1 - q)u(t - c), \tag{8.1}$$

which is not quasilinear in the up-front payment. Another example is an insurance company, which faces customers that differ in their loss probability and the degree of risk-aversion. The competitive variant of this model was first considered by Smart (2000) and Villeneuve (2003).

From an economic point of view, the main difference between nonquasilinear case and the quasilinear one is that in the former it is no longer possible speak about the efficient allocation for a given type. The reason is that the utility is now longer transferrable between a consumer of type $\boldsymbol{\alpha}$ and the monopolist. One should rather should speak of a Pareto frontier in the allocation-payment space. This means that some properties of the solution of the screening models, for example the "no distortion at the top" property, known to hold in the unidimensional quasilinear models acquire a new meaning in this context.

I start this chapter with considering the unidimensional case. I provide a complete characterization of the solution and show that the allocation and the payment of the top type lie on the Pareto frontier. Then I move to the multidimensional case and describe a scheme that allows us to check implementability of any allocation. As in the quasilinear case, the implementability conditions can be divided into the local conditions that ensure that a certain system of partial differential equations is integrable and the global condition. The global condition states the mentioned system possesses an integral, consumers' surplus, that is implementable by a tariff. It turns out that to determine whether the consumers' surplus $s(\cdot)$ is implementable by *a* tariff it is sufficient to check whether it is implementable by a particular tariff that extracts the maximum possible payment from the consumers provided that consumer of type $\boldsymbol{\alpha}$ receives surplus at least $s(\boldsymbol{\alpha})$. This result is similar to the one obtained in quasilinear case. Finally, I concentrate on the relaxed problem and provide the first order characterization of the solution.

8.1 The Unidimensional Case

Let $\Omega = (0, 1)$ and the utility of consumer of type α who obtains good of quality x and pays t is

$$u(\alpha, x, t), \tag{8.2}$$

where u is twice differentiable in all arguments,

$$u_\alpha > 0,\ u_x > 0,\ u_t < 0, u_{xx} \leq 0. \tag{8.3}$$

In some other contexts (see, for example, the model of Bardsley and Basov (2004), cited in the introduction to this chapter) one might want to interpret t as a subsidy. In that case $u_t > 0$. I will, however, assume that u_t always preserves its sign. We will say that function u satisfies the generalized Spence-Mirrlees condition if

$$\frac{\partial}{\partial \alpha}(-\frac{u_x}{u_t}) \geq 0. \tag{8.4}$$

It is a well-known fact, the if the utility satisfies the generalized Spence-Mirrlees condition (8.4) then the allocation is implementable if and only if it is increasing (see, for example, Fudenberg and Tirole, 1992). For any tariff $t(\cdot)$, let us introduce the consumer's surplus by

$$s(\alpha) = \max_x u(\alpha, x, t(x)). \tag{8.5}$$

By the envelope theorem

$$s'(\alpha) = u_\alpha(\alpha, x, t). \tag{8.6}$$

Let us for a moment concentrate on the relaxed problem. The implementability constraint can be added in the exactly the same way as in the quasilinear case.

The relaxed problem is:

$$\max \int_0^1 (t - c(x)) f(\alpha) d\alpha \tag{8.7}$$

$$s.t. s'(\alpha) = u_\alpha(\alpha, x, t), s - u(\alpha, x, t) = 0, \tag{8.8}$$

$$s(\alpha) \geq 0. \tag{8.9}$$

Here I assumed that the outside option is type independent and normalized to be zero and the types are distributed according to a strictly positive differentiable density $f(\cdot)$. Assume, further that the cost of production, $c(\cdot)$ is twice differentiable, increasing, and strictly convex. This is a standard optimal control problem with the state variables s and two control variables x and t. The Hamiltonian for this problem is:

$$H = (t - c(x))f + \lambda u_\alpha(\alpha, x, t) - \mu(s - u(\alpha, x, t)) \tag{8.10}$$

and the first order conditions are

$$\left\{ \begin{array}{c} \lambda'(\alpha) = \mu \\ f + \mu u_t + \lambda u_{\alpha t} = 0 \\ \lambda u_{\alpha x} + \mu u_x - f c'(x) = 0 \\ \lambda(1) = 0 \end{array} \right. . \tag{8.11}$$

For $\alpha = 1$ system (8.11) implies that

$$c'(x) = -\frac{u_x}{u_t}. \tag{8.12}$$

Note that the generalized Spence-Mirrlees (8.4) condition implies that $x(\cdot)$ is increasing in the neighborhood of $\alpha = 1$, therefore there is no bunching at the top. Moreover, condition (8.12) implies that allocation-transfer pair $(x(1), t(1))$ is on the Pareto frontier. Indeed, the Pareto frontier for the consumer of type α and the monopolist is the uniparametric set of solutions of the problem

$$\left\{ \begin{array}{l} \max(t - c(x)) \\ s.t. u(\alpha, x, t) \geq \overline{u} \end{array} \right. . \tag{8.13}$$

The points on the Pareto frontier are parametrized by the value of $\overline{u}$. Our assumptions ensure that the constraint in (8.13) binds and the first order conditions are necessary and sufficient. Therefore,

$$\begin{cases} 1 = -\gamma u_t \\ c'(x) = \gamma u_x \end{cases} \tag{8.14}$$

for some $\gamma > 0$, which implies (8.4). We conclude that an allocation-transfer pair is Pareto optimal if and only if (8.4) holds and hence, $(x(1), t(1))$ lies on the Pareto frontier.

Example 195 *Assume the consumer's utility function is*

$$u(\alpha; x, t) = \alpha\sqrt{x-t} - t, \tag{8.15}$$

and α is distributed uniformly on interval $(0,1)$. It is straightforward to check that utility (8.15) satisfies the generalized Spence-Mirrlees condition (8.4). Indeed,

$$\frac{\partial}{\partial\alpha}(-\frac{u_x}{u_t}) = \frac{u_t^2}{u_x^2\alpha^2\sqrt{x-t}} > 0. \tag{8.16}$$

Therefore an allocation is implementable if and only if it is increasing. The monopolist's cost of production is given by

$$c(x) = cx. \tag{8.17}$$

for some $c \in (0,1)$. For any tariff $t(\cdot)$, let us introduce the consumer's surplus by

$$s(\alpha) = \max_{x \geq 0}(\alpha\sqrt{x-t} - t) \tag{8.18}$$

and assume that the value of the outside option is type independent and normalized to be zero. Note that since $s(0) = -t \leq 0$, since the monopolist will never want to set a negative tariff, the participation constraint implies that $s(0) = 0$. Let us concentrate on the relaxed problem for the moment. Later we will see that its solution is increasing and therefore, implementable. The monopolist's problem is

$$\max \int_0^1 (t - cx)d\alpha \tag{8.19}$$

$$s.t. s'(\alpha) = \sqrt{x-t},\ s - \alpha\sqrt{x-t} + t = 0 \tag{8.20}$$

$$s(0) = 0. \tag{8.21}$$

The Hamiltonian for this problem is

$$H = t - cx + \lambda\sqrt{x-t} + \mu(\alpha\sqrt{x-t} - t - s), \tag{8.22}$$

and the first order conditions are

$$\begin{cases} \lambda' = \mu, \lambda(1) = 0 \\ s'(\alpha) = \sqrt{x-t}, s(0) = 0 \\ 1 - \frac{\lambda}{2\sqrt{x-t}} - \frac{\mu\alpha}{2\sqrt{x-t}} - \mu = 0 \\ -c + \frac{\lambda}{2\sqrt{x-t}} + \frac{\mu\alpha}{2\sqrt{x-t}} = 0 \end{cases} \tag{8.23}$$

The last two equations of this system imply that $\mu = 1 - c$ and then the first equation implies that

$$\lambda(\alpha) = (1-c)(\alpha - 1). \tag{8.24}$$

The third equation now allows us to conclude that

$$\sqrt{x-t} = \frac{1-c}{2c}(2\alpha - 1). \tag{8.25}$$

This, in turn implies that only types with $\alpha > 1/2$ are served in the equilibrium and the surplus is

$$s(\alpha) = \begin{cases} (1-c)(\alpha^2 - \alpha + 1/4)/2c, \ \textit{for}\ \alpha \geq 1/2 \\ 0, \ \textit{for}\ \alpha < 1/2 \end{cases}. \tag{8.26}$$

Finally,

$$t(\alpha) = \begin{cases} (1-c)(\alpha^2 - 1/4)/2c, \ \textit{for}\ \alpha \geq 1/2 \\ 0, \ \textit{for}\ \alpha < 1/2 \end{cases}. \tag{8.27}$$

and

$$x(\alpha) = \begin{cases} \frac{(1-c)}{2c}(\alpha^2 - 1/4) + (\frac{(1-c)(2\alpha-1)}{2c})^2, \ \textit{for}\ \alpha \geq 1/2 \\ 0, \ \textit{for}\ \alpha < 1/2 \end{cases}. \tag{8.28}$$

Note, that $x(\cdot)$ is increasing and therefore implementable. Equation (8.27) and (8.28) determine the tariff in the parametric form.

8.2 The Multidimensional Case

Let $\Omega \subset R^m_+$ be a convex, open, bounded set and the utility of consumer of type $\boldsymbol{\alpha}$ who obtains good of quality $\mathbf{x}$ and pays t is

$$u(\boldsymbol{\alpha}, \mathbf{x}, t), \tag{8.29}$$

where u is twice continuously differentiable in $\mathbf{x}$ and $\boldsymbol{\alpha}$ and both $u(\boldsymbol{\alpha}, \mathbf{x}, \cdot)$ and $\nabla_{\boldsymbol{\alpha}} u(\boldsymbol{\alpha}, \mathbf{x}, \cdot)$ are analytic.(In the quasilinear case $u(\boldsymbol{\alpha}, \mathbf{x}, \cdot)$ is affine and $\nabla_{\boldsymbol{\alpha}} u(\boldsymbol{\alpha}, \mathbf{x}, \cdot)$ is constant, therefore they are both analytic). Moreover, I assume that u is strictly increasing in the consumer's type and quality and strictly decreasing in the tariff paid. For any continuous function $\varphi(\cdot)$, let $\tau(\boldsymbol{\alpha}, \mathbf{x}, \varphi(\boldsymbol{\alpha}))$ be the unique solution of the equation

$$\varphi = u(\cdot, \cdot, \tau). \tag{8.30}$$

Definition 196 *Define the conjugate of the* $\varphi(\cdot)$ *by*

$$\varphi^*(x) = \max_{\alpha \in \overline{\Omega}} \tau(\boldsymbol{\alpha}, \mathbf{x}, \varphi(\boldsymbol{\alpha})). \tag{8.31}$$

Definition 197 *A surplus function* $s(\cdot)$ *is called implementable if there exists a measurable tariff* $t(\cdot) : R_+^n \to R \cup \{+\infty\}$ *such that*

$$s(\boldsymbol{\alpha}) = \max_{\mathbf{x} \in R_+^n} u(\boldsymbol{\alpha}, \mathbf{x}, t(\mathbf{x})) \tag{8.32}$$

for any $\boldsymbol{\alpha} \in \Omega$.

Note that this definition of the implementability differs from one we used in the previous chapters, since it concerns itself only with the surplus function rather than with the surplus and the allocation. This will allow us to separate the global incentive compatibility considerations, which rely on the generalized convexity types of arguments, from the local ones, which include the integrability conditions for a system of partial differential equations.

8.2.1 Implementability of a Surplus Function

Let us start with establishing the conditions for implementability of a surplus function. Similar to the quasilinear case, to check whether a consumers' surplus $s(\cdot)$ is implementable by *a* tariff it is sufficient to check whether it is implementable by a particular tariff, which is equal to its conjugate. This tariff, which I will call the *test tariff* has a property that it extracts the maximum possible payment from the consumers provided that consumer of type $\boldsymbol{\alpha}$ receives surplus at least $s(\boldsymbol{\alpha})$. To establish this result let us start with the following definition.

Definition 198 *Function* $s^{**}(\cdot)$ *defined by*

$$s^{**}(\boldsymbol{\alpha}) = \sup_{x \in R_+^n} u(\boldsymbol{\alpha}, \mathbf{x}, s^*(\mathbf{x})) \tag{8.33}$$

is called the biconjugate of $s(\cdot)$.

The following proposition is similar to the one proved in quasilinear case and its proof is left as an exercise to the reader.

Proposition 199 *For any continuous function* $s(\cdot)$

$$s^{**}(\boldsymbol{\alpha}) \leq s(\boldsymbol{\alpha}). \tag{8.34}$$

Now I am ready to formulate and prove the implementability criterion.

Theorem 200 *A surplus function* $s(\cdot)$ *is implementable if and only if*

$$s^{**}(\boldsymbol{\alpha}) = s(\boldsymbol{\alpha}). \tag{8.35}$$

Proof. Clearly, if (8.35) holds the surplus function is implementable and the implementing tariff is $s^*(\mathbf{x})$. Now, suppose that the surplus function is implementable, then there exists a tariff $t(\cdot) : R^n_+ \to R \cup \{+\infty\}$ such that

$$s(\boldsymbol{\alpha}) = \max_{\mathbf{x}} u(\boldsymbol{\alpha}, \mathbf{x}, t(\mathbf{x})) \tag{8.36}$$

for any $\boldsymbol{\alpha} \in \Omega$. Therefore,

$$s(\boldsymbol{\alpha}) \geq u(\boldsymbol{\alpha}, \mathbf{x}, t(\mathbf{x})) \tag{8.37}$$

for any $(\boldsymbol{\alpha}, \mathbf{x}) \in \Omega \times R^n_+$. Since $u(\boldsymbol{\alpha}, \mathbf{x}, \cdot)$ is decreasing

$$t(\mathbf{x}) \geq \boldsymbol{\tau}(\boldsymbol{\alpha}, \mathbf{x}, s(\boldsymbol{\alpha})), \tag{8.38}$$

where $\boldsymbol{\tau}(\boldsymbol{\alpha}, \mathbf{x}, s(\boldsymbol{\alpha}))$ is defined by (8.30). Therefore, taking maximum over $\boldsymbol{\alpha}$ of both parts of (8.38) one obtains:

$$t(\mathbf{x}) \geq s^*(\mathbf{x}), \tag{8.39}$$

and therefore

$$s(\boldsymbol{\alpha}) = \max_{\mathbf{x} \in R^n_+} u(\boldsymbol{\alpha}, \mathbf{x}, t(\mathbf{x})) \leq \sup_{\mathbf{x} \in R^n_+} u(\boldsymbol{\alpha}, \mathbf{x}, s^*(\mathbf{x})) = s^{**}(\boldsymbol{\alpha}) \leq s(\boldsymbol{\alpha}). \tag{8.40}$$

Here I used the fact that $u(\boldsymbol{\alpha}, \mathbf{x}, \cdot)$ is decreasing, the definition of $s^{**}(\boldsymbol{\alpha})$ and the previous proposition. Therefore,

$$s^{**}(\boldsymbol{\alpha}) = s(\boldsymbol{\alpha}). \tag{8.41}$$

■

This proposition allows us to characterize the class of implementable surpluses of a given utility. Our next objective is to characterize the set of implementable allocations.

8.2.2 *Implementability of an Allocation*

In this Section I am going to describe an algorithm that will allow us to decide whether a particular allocation is implementable. Let us start with the following definition.

Definition 201 *An allocation $x(\cdot)$ is called implementable if there exists a measurable tariff $t(\cdot) : R^n_+ \to R \cup \{+\infty\}$ such that*

$$x(\alpha) = \arg\max_{x \in R^n_+} u(\alpha, x, t(x)) \tag{8.42}$$

for any $\alpha \in \Omega$. It is called $L-$implementable if there exists a measurable tariff $t(\cdot) : R^n_+ \to R \cup \{+\infty\}$ such that(8.42) holds and

$$\int_{\Omega} (t(x) - c(x)) dF(\alpha) \geq -L. \tag{8.43}$$

The $L-$implementable allocations are introduced here for the first time. They are implementable allocations that can be implemented at a loss to of no more than L to the monopolist. Note that in the quasilinear case any implementable allocation is $L-$implementable. One has simply to augment the implementing tariff by an appropriate additive constant.

Suppose $x(\cdot)$ is an implementable allocation, $t(\cdot)$ the implementing tariff and $s(\cdot)$ the corresponding surplus. Then the generalized envelope theorem (Milgrom and Segal, 2002) applied to the consumer's optimization problem implies that

$$\nabla s(\alpha) = \frac{\partial u}{\partial \alpha_i}(\alpha, x(\alpha), \tau(\alpha, x, s(\alpha)) \equiv b_i(\alpha, s(\alpha)), \tag{8.44}$$

almost everywhere on Ω, where $\boldsymbol{\tau}(\alpha, x, s(\alpha))$ is defined by (8.30).

There are three kind of issue involved in the implementability of an allocation. First, the system of partial differential equations (8.44) should have a solution. This gives rise to the so-called *integrability problem.* Second, at least one of the solutions to (8.44) should belong to the class of the implementable surpluses characterized in the previous Section. Third, maximum in (8.42) should be achieved at the given allocation.

Let us start with discussing the integrability problem. For this purpose, define vector field

$$\mathbf{f}(\boldsymbol{\alpha}) = \mathbf{b}(\boldsymbol{\alpha}, s(\boldsymbol{\alpha})) \tag{8.45}$$

System (8.44) has a solution if and only if the vector field $f(\cdot)$ is conservative. A necessary and sufficient condition for this is (I, Theorem 4).

$$\frac{\partial f_i}{\partial \alpha_j} = \frac{\partial f_j}{\partial \alpha_i} \tag{8.46}$$

or using the definition of $f(\cdot)$ and system (8.44):

$$\frac{\partial b_i}{\partial \alpha_j} + \frac{\partial b_i}{\partial s} b_j = \frac{\partial b_j}{\partial \alpha_i} + \frac{\partial b_j}{\partial s} b_i. \tag{8.47}$$

System (8.47) is a system of $n(n-1)/2$ algebraic equation in one unknown, s. It should hold for almost all $\alpha \in \Omega$. Three things can happen: system (8.47) has no solutions, system (8.47) is satisfied identically, or it has at most countably many solutions $\{\xi^k\}_{k=0}^N$. Moreover, if $N = \infty$

$$\lim_{k \to \infty} \xi^k(\alpha) = \infty, \tag{8.48}$$

for almost all $\alpha \in \Omega$. The exhaustiveness of this list follows from the assumption of analyticity of the utility and its gradient in type. Indeed, this assumption implies that if system (8.47) is satisfied on a set with a limit point it should be satisfied identically. This is the only place, where this assumption is used.

If the first alternative holds the allocation is not implementable. Suppose the second alternative holds. Then the system has a one parametric set of solutions that can be found in the following way. First, consider equation

$$\frac{\partial s}{\partial \alpha_1} = b_1(\alpha, s) \tag{8.49}$$

and treat variables $\alpha_2, ..., \alpha_m$ as parameters. Then it is a first order ordinary differential equation. Therefore, it has a uniparametric family of solutions

$$s = \Psi(\alpha_1, C), \tag{8.50}$$

where

$$C = \varphi(\alpha_2, ..., \alpha_m). \tag{8.51}$$

Substitute (8.50)-(8.51) into

$$\frac{\partial s}{\partial \alpha_2} = b_2(\alpha, s) \tag{8.52}$$

to obtain

$$\frac{\partial \varphi}{\partial \alpha_2} = \frac{b_2(\alpha, \Psi(\alpha, \varphi))}{\Psi_\varphi(\alpha, \varphi)}. \tag{8.53}$$

Note that the right hand side of (8.53) does not depend on α_1. Indeed, consider $\alpha_3, ..., \alpha_m$ as parameters and fix them at some values. Then,

$$\frac{\partial}{\partial \alpha_1}\left(\frac{b_2(\alpha, \Psi(\alpha, \varphi))}{\Psi_\varphi(\alpha, \varphi)}\right) = \frac{1}{\Psi_\varphi^2}\left(\left(\frac{\partial b_2}{\partial \alpha_1} + \frac{\partial b_2}{\partial s} b_1\right)\Psi_\varphi - \frac{\partial^2 \Psi}{\partial \varphi \partial \alpha_1} b_2\right). \tag{8.54}$$

But

$$b_2 \frac{\partial^2 \Psi}{\partial \varphi \partial \alpha_1} = b_2 \frac{\partial}{\partial \varphi}(b_1(\alpha, \Psi(\alpha, \varphi)) = \left(\frac{\partial b_1}{\partial \alpha_2} + \frac{\partial b_1}{\partial s} b_2\right)\Psi_\varphi. \tag{8.55}$$

The last equality uses the fact that since $\alpha_3, ..., \alpha_m$ are assumed to be fixed, and therefore

$$\frac{\partial \alpha_2}{\partial \varphi} = \frac{\Psi_\varphi}{b_2}. \tag{8.56}$$

Now, using conditions (8.47) we conclude that the right hand side of (8.53) does not depend on α_1 and we succeeded in eliminating variable α_1 from (8.44). Proceeding in a similar way we can solve the system (8.44) recursively to obtain a uniparametric family of solutions

$$s = s(\alpha, c), \tag{8.57}$$

Note that though (8.44) is a system of partial differential equations at each step we had to solve a first order ordinary differential equation.

Having found the solution to (8.44) define the family of the test tariffs

$$s^*(x,c) = \max_{\alpha\in\Omega} \tau(\alpha, x, s(\alpha, c)). \tag{8.58}$$

Then allocation $x(\alpha)$ is implementable if and only if

$$x(\alpha) \in \arg\max u(\alpha, x, s^*(x,c)) \tag{8.59}$$

for some $c \in R$.

Finally, suppose that the third alternative holds. Select a solution $\xi^j(\boldsymbol{\alpha})$ of conditions (8.44) and check whether the system

$$\nabla\xi^j(\alpha) = b_i(\alpha, \xi_j(\alpha)). \tag{8.60}$$

is satisfied. If the system is satisied for no j the allocation is not implementable. On the other hand, if for some $j = k$ equation (8.60) holds, define the test tariff by

$$\xi_k^*(x) = \max_{\alpha\in\Omega} \tau(\alpha, x, \xi_k(\alpha)) \tag{8.61}$$

and check whether

$$x(\alpha) \in \arg\max u(\alpha, x, \xi_k^*(x)). \tag{8.62}$$

The allocation is implementable if and only if (8.62) holds for some k. Note that if one is interested in $L-$implementable allocations for some finite L, i. e. allocations that can be implemented without imposing loss bigger than L on the monopolist, one has to check only a finite set of solutions of (8.44) according to (8.48). Anyway, the situation with countable set of solutions is not typical.

The results of this Section can be summarized in the following theorem:

Theorem 202 *For allocation $x(\cdot)$ to be implementable the system (8.44) should be compatible. If the latter is the case, $x(\cdot)$ is implementable if and only if it is implementable by tariff $s^*(x)$, where $s(x)$ is a solution to (8.44).*

8.3 The First Order Conditions for the Relaxed Problem

The relaxed screening problem is:

$$\max \int_\Omega (t - c(\mathbf{x}))f(\boldsymbol{\alpha})\mathbf{d}\boldsymbol{\alpha} \tag{8.63}$$

$$\begin{aligned} s.t.\nabla s(\boldsymbol{\alpha}) &= \nabla_{\boldsymbol{\alpha}} u(\boldsymbol{\alpha}, \mathbf{x}, t), && (8.64)\\ s-u(\boldsymbol{\alpha}, \mathbf{x}, t) &= 0, && (8.65)\\ s(\boldsymbol{\alpha}) &\geq 0. && (8.66)\end{aligned}$$

Here I assumed that the outside option is type independent and normalized to be zero and the types are distributed according to a strictly positive differentiable density $f(\cdot)$. The problem is completely similar to the quasi-linear case, with the only difference that there exists and additional control variable t and an additional constraint (8.65). Let us for simplicity assume that $n \geq m$. Otherwise, we have to introduce the utils in the same way as in Chapter 7 and consider additional phase constraints that take into account that not all utils combinations are feasible.

Let distribution η be the Lagrange multiplier on the participation constraint. Then the relaxed problem is equivalent to

$$\begin{cases} \max \int_\Omega [(t - c(\mathbf{x}))f(\boldsymbol{\alpha}) + \eta(\boldsymbol{\alpha})s(\boldsymbol{\alpha})]d\boldsymbol{\alpha} \\ \quad s.t. \nabla s(\boldsymbol{\alpha}) = \nabla_{\boldsymbol{\alpha}} u(\boldsymbol{\alpha}, \mathbf{x}, t) \\ \quad s - u(\boldsymbol{\alpha}, \mathbf{x}, t) = 0 \end{cases} . \tag{8.67}$$

Let us make the following assumption.

Condition 203 *Problem*

$$V(\boldsymbol{\alpha}, a, \mathbf{b}) = \max_{(\mathbf{x},t) \in R^n_+ \times R_+} [(t - c(\mathbf{x}))f(\boldsymbol{\alpha}))] \tag{8.68}$$

$$s.t. \mathbf{b} = \nabla_{\boldsymbol{\alpha}} u(\boldsymbol{\alpha}, \mathbf{x}, t) \tag{8.69}$$

$$u(\boldsymbol{\alpha}, \mathbf{x}, t) = a. \tag{8.70}$$

has a solution for any $a \in R$, $\mathbf{b} \in R^m_+$.

The relaxed problem can be solved in two steps. First, let $s \in H^1(\Omega)$ be any surplus function. Define the value function

$$V(\boldsymbol{\alpha}, s, \nabla s) = \max_{\mathbf{x},t} [(t - c(\mathbf{x}))f(\boldsymbol{\alpha}))] \tag{8.71}$$

$$s.t. \nabla s(\boldsymbol{\alpha}) = \nabla_{\boldsymbol{\alpha}} u(\boldsymbol{\alpha}, \mathbf{x}, t) \tag{8.72}$$

$$u(\boldsymbol{\alpha}, \mathbf{x}, t) = s. \tag{8.73}$$

Let $\mathbf{x}(\boldsymbol{\alpha}, s, \nabla s)$ and $t(\boldsymbol{\alpha}, s, \nabla s)$ solve (8.71)-(8.73). According to Condition 204 the solution exists. Now select the surplus to solve

$$\max \int_\Omega (V(\boldsymbol{\alpha}, s, \nabla s) + \eta(\boldsymbol{\alpha})s(\boldsymbol{\alpha}))d\boldsymbol{\alpha}. \tag{8.74}$$

If s^* solves (8.74) then $\mathbf{x}(\boldsymbol{\alpha}, s^*, \nabla s^*)$ and $t(\boldsymbol{\alpha}, s^*, \nabla s^*)$ solve the relaxed problem. According to the generalized envelope theorem (I, Section 5.9) the value function is almost everywhere differentiable and satisfies the envelope conditions. Therefore, the Euler-Lagrange equation for problem (8.74) implies

$$\sum_{j=1}^{m} \frac{\partial}{\partial \alpha_j} \left(\frac{\partial V}{\partial s_{\alpha_j}}\right) = \frac{\partial V}{\partial s}, \tag{8.75}$$

and the free boundary condition states that

$$\sum_{i=1}^{m} n_i \frac{\partial V}{\partial s_{\alpha_j}} = 0 \text{ on } \partial\Omega. \tag{8.76}$$

To evaluate the derivatives in (8.75) form the Lagrangian for problem (8.71)-(8.73) :

$$L = (t - c(\mathbf{x}))f(\boldsymbol{\alpha}) + \eta(\boldsymbol{\alpha})s(\boldsymbol{\alpha}) + \boldsymbol{\lambda}\cdot\nabla_{\boldsymbol{\alpha}}u(\boldsymbol{\alpha},\mathbf{x},t) + \mu(s - u(\boldsymbol{\alpha},\mathbf{x},t)), \tag{8.77}$$

Then by the envelope theorem for constraint maximization:

$$\frac{\partial V}{\partial s} = \frac{\partial L}{\partial s} = \eta + \mu, \tag{8.78}$$

$$\frac{\partial V}{\partial s_{\alpha_j}} = \frac{\partial L}{\partial s_{\alpha_j}} = -\lambda_j. \tag{8.79}$$

The results can be summarized in the following theorem:

Theorem 204 *Let an allocation* $\mathbf{x}(\cdot)$, *and tariff* $t(\cdot)$ *solve the relaxed problem. Then there exist distributions* η, μ *a vector function* $\boldsymbol{\lambda} \in H^1(\Omega)$ *and a function* $s \in H^1(\Omega)$ *such that the following system is satisfied:*

$$div\boldsymbol{\lambda} = -\frac{\partial H}{\partial s} \tag{8.80}$$

$$\boldsymbol{\lambda}\cdot\mathbf{n} = 0 \text{ on } \partial\Omega \tag{8.81}$$

$$\nabla s(\boldsymbol{\alpha}) = \nabla_{\boldsymbol{\alpha}}u(\boldsymbol{\alpha},\mathbf{x},t) \tag{8.82}$$

$$s - u(\boldsymbol{\alpha},\mathbf{x},t) = 0 \tag{8.83}$$

$$s \geq 0,\ \eta \geq 0,\ \eta s = 0 \tag{8.84}$$

$$(\mathbf{x},t) \in \arg\max H(\mathbf{x},t,s;\boldsymbol{\lambda},\eta,\mu), \tag{8.85}$$

where the Hamiltonian H is given by

$$H = (t - c(\mathbf{x}))f(\boldsymbol{\alpha}) + \eta(\boldsymbol{\alpha})s(\boldsymbol{\alpha}) + \boldsymbol{\lambda}\cdot\nabla_{\boldsymbol{\alpha}}u(\boldsymbol{\alpha},\mathbf{x},t) + \mu(s - u(\boldsymbol{\alpha},\mathbf{x},t)). \tag{8.86}$$

Note that though this Theorem sounds similar to Theorem 192, the crucial difference is that here we do not know whether the solution exists even for the relaxed problem. This is why we had to impose Condition 204. As we will see in the next chapter, in the quasilinear case the existence of the solution to the relaxed problem can be proved under rather general assumptions. The conditions under which the existence of the solution to the complete problem is established remain somewhat more restrictive.

8.4 Exercises

1. Prove inequality (8.34).
2. Let

$$u(\boldsymbol{\alpha}, \mathbf{x}) = -\alpha_1 \exp(t - x_1) - \alpha_2 \exp(t - x_2). \tag{8.87}$$

Prove that no allocation such that $x_1(\boldsymbol{\alpha}) \equiv x_2(\boldsymbol{\alpha}) \neq 0$ is implementable.

8.5 Bibliographic Notes

The generalized Spence-Mirrlees condition for the unidimensional case is well known and can be found, for example, in Fudenberg and Tirole (1992). However, to the best of my knowledge there are no examples of even unidimensional non-quasilinear problems solved in the literature. All results on the multidimensional non-quasilinear models a new and were not published anywhere. Bardsley and Basov (2004) provided the first example of relevance of such models in practical economic situations. The ideas that were used to derive the frist order characterization of the relaxed problem were first used in the context of one-dimensional optimal control problems by Long and Shimomura (2003).

9

Existence, Uniqueness, and Continuity of the Solution

So far we derived the first order characterization of the solution of the screening models. We, however, left aside the question whether the solution to the problem exists. In this chapter we are going to consider existence and uniqueness of the solution for both the relaxed and the complete problem.

9.1 Existence and Uniqueness for the Relaxed Problem

Let us consider the relaxed screening problem. The following theorem establishes existence and uniqueness of the solution.

Theorem 205 *Let us assume that*

(i) there exists $\gamma > 0$ such that $f(\alpha) \geq \gamma$ for any $\alpha \in \Omega$;

(ii) the functions $\partial u / \partial \alpha_i(\alpha, \cdot) : R_+^n \to R$ are twice continuously differentiable and concave;

(iii) $c(\cdot)$ is twice continuously differentiable and the matrix of its second derivatives $B = D^2 c$ has uniformly bounded eigenvalues, i. e. $\exists \varepsilon > 0$ and $\exists M > \varepsilon$ such that $\forall x \in R_+^n$ and $\forall h \in R^n$

$$\varepsilon \|h\|^2 \leq \langle h, Bh \rangle \leq M \|h\|^2 .$$

Then the relaxed problem has a unique solution.

Proof. Define functional π by a formula:

$$\pi(s) = \int_\Omega (u(a,z) - c(z) - s(\alpha))f(\alpha)d\alpha. \qquad (9.1)$$

Let

$$\begin{aligned} K &= \{s \in H^1(\Omega) : \frac{\partial s}{\partial \alpha_i}(\alpha) = \frac{\partial u}{\partial \alpha_i}(\alpha, z'),\ i = \overline{1,m}, s(\alpha) \geq s_0(\alpha), \\ z_i &\leq \frac{\partial u}{\partial \alpha_i}(\alpha, z'),\ i = \overline{1, m-n}\}. \end{aligned} \qquad (9.2)$$

I will prove that there exists a unique surplus function $s^* \in K$ and allocation $z^*(\cdot)$ such that $\pi(s^*) \geq \pi(s)$ for any $s \in K$ and $z_i^* = \partial v / \partial \alpha_i(\alpha, z^{*\prime})$. Since under assumptions on functions $\partial v / \partial \alpha_i$ the set K is a convex closed set and the functional $\pi(\cdot)$ is strictly concave. Therefore, to prove the existence it is sufficient to prove that the functional π is coercive on K, i. e. that $\pi(s)$ tends to $-\infty$ when $|s|_{C_0^\infty(R^n)^1}$ tends to $+\infty$. For all $s \in H^1(\Omega)$, denote by $\underline{s}$ the mean value of s over Ω:

$$\underline{s} = \frac{1}{|\Omega|} \int_\Omega s(\alpha) d\alpha. \qquad (9.3)$$

By the Poincare inequality there exists a constant $M(\Omega)$ such that for all $s \in H^1(\Omega)$, $|s - \underline{s}|_{L^2} \leq M(\Omega) |\nabla s|_{L^2}$. This implies that

$$|s|_{H^1} \to +\infty \Leftrightarrow \underline{s} \to +\infty \text{ or } |\nabla s|_{L^2} \to +\infty\ . \qquad (9.4)$$

Note that since each $v_j(\cdot)$ is concave and $v_j(0) = 0$ one obtains $z \leq \langle \nabla v(0), z' \rangle$ and hence $|\nabla s|_{L^2} \leq N(\Omega) |z'|_{L^2}$ .Under the assumptions on the cost and the distribution of types:

$$\pi(s) \leq -\gamma\varepsilon |z|_{L^2}^2 + M(\Omega)N(\Omega) |z|_{L^2} - \underline{s}. \qquad (9.5)$$

Note that since $z_j \leq u_j(z')$ for $\overline{1, m-n}$ the condition $|\nabla s|_{L^2} \to +\infty$ implies $z' \to +\infty$. Coerciveness of π then follows from (9.4) and (9.5). Hence, I have proven that π achieves maximum for some $s^* \in K$. To see that $z_i^* = \partial v / \partial \alpha_i(\alpha, z^{*\prime})$ should hold, assume that there exists k such that $z_k < \partial v / \partial \alpha_k(\alpha, z')$. Consider a function

$$s'(\alpha) = s^*(\alpha) + \delta\alpha_i \ - \varepsilon \qquad (9.6)$$

for each α such that $s^*(\alpha) > s_0(\alpha)$ and $i = \overline{1, m-n}$. Note that $s' \in K$ for sufficiently small $\delta > 0$ and $\varepsilon > 0$. Since the cost function depends only on z', the integrand in the definition of π increases by ε. If $0 \notin \Gamma$, one can find such values of δ and ε such that new participation region is a superset of the initial one. Otherwise, some points may drop out of the participation region, but their Lebesgue measure will be $O(\varepsilon^m)$. In any case, $\pi(s') > \pi(s^*)$ for sufficiently small δ and ε. This implies that $z_i^* = \partial v/\partial \alpha_i(\alpha, z^{*\prime})$, which completes the proof of the existence.

Let $s_1, s_2 \in K$ both solve

$$\min \pi(s) \tag{9.7}$$

$$s.t. s \in K. \tag{9.8}$$

Since K is convex $s = 1/2(s_1 + s_2) \in K$ and since functional $\pi(\cdot)$ is strictly concave

$$\pi(s) > \pi(s_1). \tag{9.9}$$

Therefore, the maximizer is unique. ■

In the case $m = n$ and

$$u(\boldsymbol{\alpha}, \mathbf{x}) = \sum_{i=1}^{n} \alpha_i x_i \tag{9.10}$$

Rochet and Chone (1998) proved that the allocation is continuous. In that case the above proof can be used also verbatim to demonstrate the existence and the uniqueness of the solution to the *complete* problem, since set of the implementable surpluses is still a convex closed set. The same is true for the linear problems with arbitrary m and n (Basov, 2001). One has only to notice that if $s^*(\cdot)$, the solution to the problem (9.7)-(9.8) is convex so is $s'(\cdot)$ defined by (9.6). The existence of the solution for the complete problem in the general case is a more delicate issue.

9.2 Existence of a Solution for the Complete Problem

The existence of the solution of the complete problem under some general conditions is still an open problem. The first step in this direction was undertaken by Carlier (2001) but the assumptions are too restrictive to make this result really interesting. The advantage of the result, however, is that when applicable it can provide conditions for existence beyond the quasilinear case. Another approach is taken by Monteiro and Page (1998). They assumed that the consumers are budget constraint and consumer the monopolist's good as well as a bundle of goods purchased on a outside competitive market. They proved that the existence can be guaranteed for all nonlinear pricing problems if and only if the monopolist's good is nonessential, i. e. consumer can derive utility form other goods without consuming the monopolist's good but cannot derive utility from the monopolist's good without other goods. This last assumptions sets this paper aside from the rest of the literature in the area. The important advantage of their approach is that they allow the type space to be an arbitrary measure space, rather than a subset of a Euclidean space. The price is that they did not go beyond proving the existence and did not provide any characterization of the solution.

9.3 Continuity of the Solution

The first result concerning the continuity of the solution was obtained by Mussa and Rosen (1978). They considered a very specific utility function

$$u(\alpha, x) = \alpha x. \tag{9.11}$$

Here I will present a slightly generalized version of their result.

Theorem 206 *Assume that the types distribution $f(\cdot)$ is continuous and $f(\alpha) > 0$ for any $\alpha \in \Omega$, the consumers' $u(\cdot)$ is twice differentiable, satisfies the Spence-Mirrlees condition ($u_{12} > 0$),*

$$MR(\alpha, x) \equiv u_2(\alpha, x) - \frac{1 - F(\alpha)}{f(\alpha)} u_1(\alpha, x) \tag{9.12}$$

is decreasing in x, and the cost function $c(\cdot)$ is twice differentiable and strictly convex. Then the optimal allocation, $x^*(\cdot)$, is continuous.

Proof. Recall from Chapter 6 that the monopolist's problem is

$$\max \int_0^1 (u(\alpha, x) - s(\alpha) - c(x)) f(\alpha) d\alpha \tag{9.13}$$

$$\begin{aligned} \text{s.t. } s'(\alpha) &= u_1(\alpha, x), && (9.14) \\ x'(\alpha) &= \gamma, && (9.15) \\ \gamma &\geq 0, && (9.16) \\ s(0) &= 0. && (9.17) \end{aligned}$$

The corresponding Hamiltonian is

$$H(\alpha, \gamma; x, s, \lambda_1, \lambda_2) = (u(\alpha, x) - s(\alpha) - c(x)) f(\alpha) + \lambda_1 u_1(\alpha, x) + \lambda_2 \gamma + \mu \gamma$$

and the first order conditions imply:

$$\begin{aligned} \lambda_1'(\alpha) &= f(\alpha) && (9.18) \\ \lambda_2'(\alpha) &= -[(u_2(\alpha, x) - c'(x)) f(\alpha) + \lambda_1 u_{12}(\alpha, x)] && (9.19) \\ \lambda_2 + \mu &= 0 && (9.20) \\ \lambda_1(1) &= 0, \ \lambda_2(1) = 0 && (9.21) \\ \gamma &\geq 0, \ \mu \geq 0, \ \mu\gamma = 0. && (9.22) \end{aligned}$$

Assume that the Theorem does not hold. Since $x^*(\cdot)$ is increasing, the only kind of discontinuity it can display is a jump, i. e. $\exists\alpha^* \in (0,1)$ such that

$$a = \lim_{\alpha\to\alpha^*-0} x(\alpha) < \lim_{\alpha\to\alpha^*+0} x(\alpha) = b.$$

Then

$$\gamma = (b-a)\delta(\alpha-\alpha^*) \neq 0, \tag{9.23}$$

therefore

$$\lambda_2(\alpha^*) = \mu(\alpha^*) = 0. \tag{9.24}$$

But then equations (9.18)-(9.21) imply:

$$\mu(\alpha) = -\lambda_2(\alpha) = \int_{\alpha}^{\alpha^*} [MR(\alpha,x) - c'(x)]f(\alpha)d\alpha. \tag{9.25}$$

Our assumptions on the cost and the marginal revenue functions guarantee that

$$\lim_{\alpha\to\alpha^*-0}[MR(\alpha,x(\alpha)) - c'(x(\alpha))] > \lim_{\alpha\to\alpha^*+0}[MR(\alpha,x(\alpha)) - c'(x(\alpha))]. \tag{9.26}$$

Therefore, either

$$g(\alpha) = MR(\alpha,x(\alpha)) - c'(x(\alpha)) > 0 \tag{9.27}$$

in a sufficiently small neighborhood left of α^* or

$$g(\alpha) = MR(\alpha,x(\alpha)) - c'(x(\alpha)) < 0 \tag{9.28}$$

in a sufficiently small neighborhood right of α^*. Since

$$0 = \mu(\alpha^*) = \mu(\alpha) + \int_{\alpha}^{\alpha^*} g(\beta)d\beta \tag{9.29}$$

and the last integral is positive for α sufficiently close to α^*, we conclude that $\exists\alpha \in (0,1)$ such that

$$\mu(\alpha) < 0, \tag{9.30}$$

which contradicts (9.22). ■

Intuitively, the reason for the continuity of the optimal allocation is the following: suppose the allocation jumps at α^*. Then if the monopolist decreases the amount for the types just above α^* by sufficiently small $\varepsilon > 0$ the resulting allocation will still be increasing and therefore implementable. If the point α^* is in the interior of the participation region this will increase monopolist's profits, since it will not change the participation decision and save the costs of production. Otherwise, since the quality produced by the monopolist is always below the efficient quality it is always possible to decrease payment on such amount that the same types will participate but the decrease in the revenue is still outweighed by the cost savings. Rochet and Chone (1998) generalized the result for the case with $m = n$ and utility linear in types. They used it to argue that in that case the optimal product line is always connected. Sufficient conditions for the continuity in the general case are not known. The main difficulties are expected in the case $m > n$, since in this case the monopolist does not have enough instruments to vary the utils independently.

9.4 Bibliographic Notes

The first results concerning the existence, uniqueness, and continuity of the solution were obtained in the unidimensional case by Mussa and Rosen (1978). The first existence proof for both relaxed and complete problems was obtained by Rochet and Chone (1998) in the linear case with $m = n$. The result was generalized by Basov (2001) for both relaxed and complete problems in the linear case and for the relaxed problem in the general case. The first results concerning the existence of the complete problem in the general case were obtained by Carlier (2001). His proof is based on the considerations of the optimal control problem under the generalized convexity constraint and some properties of subdifferentials of the generalized convex functions. Unfortunately, the assumption made are still too restrictive to make the results interesting for applications. Approach of Monteiro and Page (1998) is based on the methods of real analysis and allows for more general type spaces. However, not much beyond existence is known for this case.

10
Conclusions

Our understanding of the economics and mathematics of the screening models has advanced greatly during the last two decades. However, the solutions to these models are usually characterized by a complicated systems of nonlinear partial differential equation. Moreover, beyond the unidimensional case one can provide very little economic intuition for the properties of the solution. It is worth mentioning that simple solutions were obtained in some limit cases. For example, in the box demand model with one good and unknown intercept and the slope parameter the optimal tariff becomes piece-wise affine as the demand becomes perfectly elastic. This example is presented in Chapter 7 of this book. Another interesting limit case was considered by Armstrong (1999). He assumed that the consumers' utility has a form

$$U(\boldsymbol{\alpha}, \mathbf{x}, t) = \sum_{i=1}^{n} u_i(\alpha_i, x_i) - t \tag{10.1}$$

and the cost in linear in quality characteristics and proved that if the tastes for different goods are uncorrelated a two part tariff will asymptotically extract all consumer surplus as $n \to \infty$. If the tastes are correlated the job can be done by a menu of two part tariffs.

In the Armstrong's example the dimensionality of the consumer type always equals the number of the quality characteristics. Probably a more economically relevant case is to fix the number of goods and allow the dimensionality of type to go to infinity. Indeed, since from an economic point of view the consumer's type is simply her utility function and we do not want to impose any a priori restriction on tastes, apart from probably some general assumption of monotonicity and convexity, an infinite dimensional type seems a reasonable economic assumption. From the technical point of view an infinite dimensional type can be introduced in two different ways: take the limit $m \to \infty$ in the equations we obtained in this book or formulate the problem from the start in the infinite dimensional Banach space. The latter is more mathematically elegant but it will raise some new questions like what kind of integral to use (Bochner integral seems a natural candidate). I am not aware of any results in that area. But this seems an interesting area for future research.

Another interesting extension is to consider oligopolistic models with private information. Champsuar and Rochet (1989) developed a model of duopoly when the private information is one-dimensional. They stressed the important role played by the "no bunching" condition for the monopolistic model in ensuring the existence of the equilibrium in the duopoly game. In the multidimensional case, as we have seen, the prevalence of bunching is sensitive to modelling assumptions. Therefore, generalization of the model to cover the oligopolistic case will present a considerable challenge.

11
References

T. M. Abasov, and A. M. Rubinov,1994, On the class of H-convex functions, Russian Academy of Sciences Doklady Mathematics, 48, 95-97.
R. Abraham, J.E. Mardsen, and T. Ratium, 1988, Manifolds, tensor analysis, and applications, (New York: Spinger-Verlag).
Adams, W. J., and J. L. Yellen, 1976, Commodity bundling and the burden of monopoly, Quarterly Journal of Economics, 90,. 475-498
M. Armstrong, 1996, Multiproduct nonlinear pricing, Econometrica 64, 51-75.
M. Armstrong, 1999, Price discrimination by a many-product firm, Review of Economic Studies 66, 151-168.
D. Baron, and R. Myerson, 1982, Regulating a monopolist with unknown cost, Econometrica, 50, 911-930.
J. Barros-Netto, 1973, An introduction to the theory of distributions, (New York: M. Dekker).
P. Bardsley, and S. Basov, 2003, Cost of cheating and grant distribution, The University of Melbourne, Department of Economics, mimeo.
P. Bardsley, and S. Basov, 2004, A model of grant distribution: screening approach, The University of Melbourne, Department of Economics, mimeo.
S. Basov, 2001, Hamiltonian approach to multidimensional screening, Journal of Mathematical Economics, 36, 77-94.
S. Basov, 2002, A partial characterization of the solution of the multidimensional screening problem with nonlinear preferences, The University of Melbourne, Department of Economics, Research Paper #860.

S. Basov, 2004, Three approaches to multidimensional screening, in A. Tavidze, ed. Progress in Economics Research, volume 7, pp. 159-178, (New York: Nova Science Publishers).
B. J. Cantwell, 2002, Introduction to symmetry analysis, (Cambridge: Cambridge University Press).
G. Carlier, 2001, A general existence result for the principal-agent problem with adverse selection, Journal of Mathematical Economics, 35, 129-150.
G. Carlier, 2002, Duality and existence for a class of mass transportation problems and economic applications, unpublished manuscript.
www.sci.unich.it/~scarsini/erice/pdffiles/CarlierErice.pdf
M. Carter, 2001, Foundation of Mathematical Economics, (Cambridge, MA: MIT Press).
P. Champsuar, and J.-C. Rochet, 1989, Multi-product duopolies, Econometrica, 57, 533-557.
R. Courant and D. Hilbert, 1989, Methods of mathematical physics, (New York: Interscience Publishers).
C. Dellacherie, and P.-A. Meyer, 1978, Probabilities and potential, (Amsterdam: Elsevier).
Z. Eberhard, 1984, Nonlinear Functional Analysis and its Applications, III. Variational Methods and Optimization. (Berlin: Springer-Verlag)..
Eggleston, 1958, Convexity, (Cambridge: Cambridge University Press).
D. Fudenberg, and J. Tirole, 1992, Game theory, (Cambridge, MA: MIT Press).
I. M.Gelfand, and S. V. Fomin, 1963, Calculus of variations, (Englewood Cliffs, NJ: Prentice-Hall).
P. Jehiel, B. Moldovanu, and E. Stacchetti, 1999, multidimensional mechanism design for auctions with externalities, Journal of Economic Theory, 85, 258-293.
B. Jullien, 2000, Participation constraints in adverse selection models, Journal of Economic Theory, 93, 1-47.
D. Kinderlehrer, and G. Stampacchia, 1980, An introduction to variational inequalities, (Boston, MA: Academic Press).
Kolmogorov, A. N. and S. V. Fomin, 1970, Introductory real analysis, (Englewood Cliffs, NJ: Prentice-Hall).
Laffont, J. J., E. Maskin, and J. C. Rochet, 1987, Optimal nonlinear pricing with two-dimensional characteristics, in: T. Groves, R. Radner, and S. Reiter, eds., Information, incentives, and economic mechanisms, pp. 256-266, (Minneapolis: University of Minnesota Press).
F. L. Lewis, 1986, Optimal control, (New York : Wiley).
T. Lewis, and D. Sappington, 1989, Countervailing incentives in agency theory, Journal of Economic Theory 49, 294-313.
J. L. Lions, 1971, Optimal control of systems governed by partial differential equations, (Berlin : Springer-Verlag).
P. L. Lions, 1998, Identification du cône dual des fonctions convexes et applications, C.R. Acad. Sci. Paris, Serie 1, 326, 1385-1390.

N. V. Long, and K. Shimomura, 2003, A new proof of the maximum principle, Economic Theory, 22, 671-674.
D. G. Luenberger, 1969, Optimization by vector space methods, (New York: John Wiley and Sons).
A. Mas-Colell, M. D. Whinston, and J. R. Green, 1995, Microeconomic theory, (Oxford: Oxford University Press).
E. Maskin, and J. Riley, 1984, Monopoly with incomplete information, Rand Journal of Economics 15 171-196.
R. P. McAfee, and J. McMillan, 1988, Multidimensional incentive compatibility and mechanism design, Journal of Economic Theory, 46, 335-364.
P-A. Meyer, 1966, Probability and potentials, (Waltham, MA: Blaisdell Publishing Company).
P. Milgrom, and I. Segal, 2002, Envelope theorems for arbitrary choice sets, Econometrica, 70, 583-601.
J. Mirrlees, 1971, An exploration in the theory of optimum income taxation, Review of Economic Studies, 38, 175-208.
L. J. Mirman, and D. Sibley, 1980, Optimal nonlinear prices for multiproduct monopolies, Bell Journal of Economics, 11, 659-675.
M. Mussa, and S. Rosen, 1978, Monopoly and product quality, Journal of Economic Theory 18, 301-317.
P. K. Monteiro, and F. H. Page, 1998, Optimal selling mechanisms for multi-product monopolists: incentive compatibility in the presence of budget constraints, Journal of Mathematical Economics, 30, 473-502.
J. C. Rochet, 1985, The taxation principle and multitime Hamilton-Jacobi equations, Journal of Mathematical Economics 14, 113-128.
J. C. Rochet, 1987, A necessary and sufficient condition for rationalizability in a quasi-linear context, Journal of Mathematical Economics 16, 191-200.
J. C. Rochet, and P.Chone, 1998, Ironing, sweeping and multidimensional screening, Econometrica 66, 783-826.
J. C. Rochet, and L. A. Stole, 2001, Nonlinear pricing with random participation, *gsblas.uchicago.edu/papers/randpart.pdf*
J. C. Rochet, and L. A. Stole, 2003, The economics of multidimensional screening, in M. Dewatripont, L. P. Hansen, and S. J. Turnovsky, eds. Advances in economics and econometrics, Cambridge, UK: The Press Syndicate of the University of Cambridge, 115-150.
T. R. Rockafellar, 1997, Convex analysis, Princeton University Press.
H.L. Royden, 1988, Real analysis, New York: Macmillan.
A.M. Rubinov, 2002, Abstract Convexity and Global Optimization, (Boston, MA: Kluwer Academic Publishers).
Rudin, W, 1964, Principles of mathematical analysis, (New York: McGraw-Hill).
R. Sah, and J. Zhao, 1998, Some envelope theorems for integer and discrete choice variables, International Economic Review, 39, 623-634.
D. Sappington, 1983, Optimal regulation of the monopolist with unknown technological capabilities, Bell Journal of Economics, 15, 453-463.

R. Sato, and R. V. Ramachandran, 1990, Conservation laws and symmetry: applications to economics and finance, (Boston, MA : Kluwer Academic Publishers).
I. Seddon, 1957, Elements of partial differential equations, (New York: McGraw-Hill Book Company).
I. Singer, 1997, Abstract convex analysis, (New York: John Wiley and Sons).
V. I. Smirnov, 1964, A course is higher mathematics, volume II, Advanced calculus, (Oxford: Pergamon Press).
J. Shapiro, 2001, Income maintenance programs and multidimensional screening, *www.econ.upf.es/~shapiro/IMP-J.Shapiro.pdf*
M. Smart, 2000, Competitive insurance markets with two unobservable, International Economic Review, 41, 153-169.
L. A. Stole, 2000, Lectures on contracts and organizations, *http://gsblas.uchicago.edu/papers/lectures.pdf*
V. M. Tikhomirov, and A. D. Ioffe, 1979, Theory of extremal problems, (Amsterdam: North Holland Publishing Company).
A. N. Tikhonov, and A.A. Samarski, 1964, Partial differential equations of mathematical physics, (San Francisco: Holden-Day).
D. M. Topkis, 1998, Supermodularity and complementarity, (Chichester NJ: Princeton University Press).
V. S. Vladimirov, 2002, Methods of the theory of generalized functions, (New York: Taylor and Francis).
B. Villeneuve, 2003, Concurrence et antisélection multidimensionnelle en assurance, Annals d'Economie et de Statistique, 69, 119-142
R. Wilson, 1993, Non linear pricing, (Oxford: Oxford University Press).
D. Zwillinger (editor), 2003, CRC standard mathematical tables and formulae, (Boca Raton, FLA: CRC Press).